Prepared in cooperation with the Harris-Galveston Subsidence District

Investigation of Land Subsidence in the Houston-Galveston Region of Texas By Using the Global Positioning System and Interferometric Synthetic Aperture Radar, 1993–2000

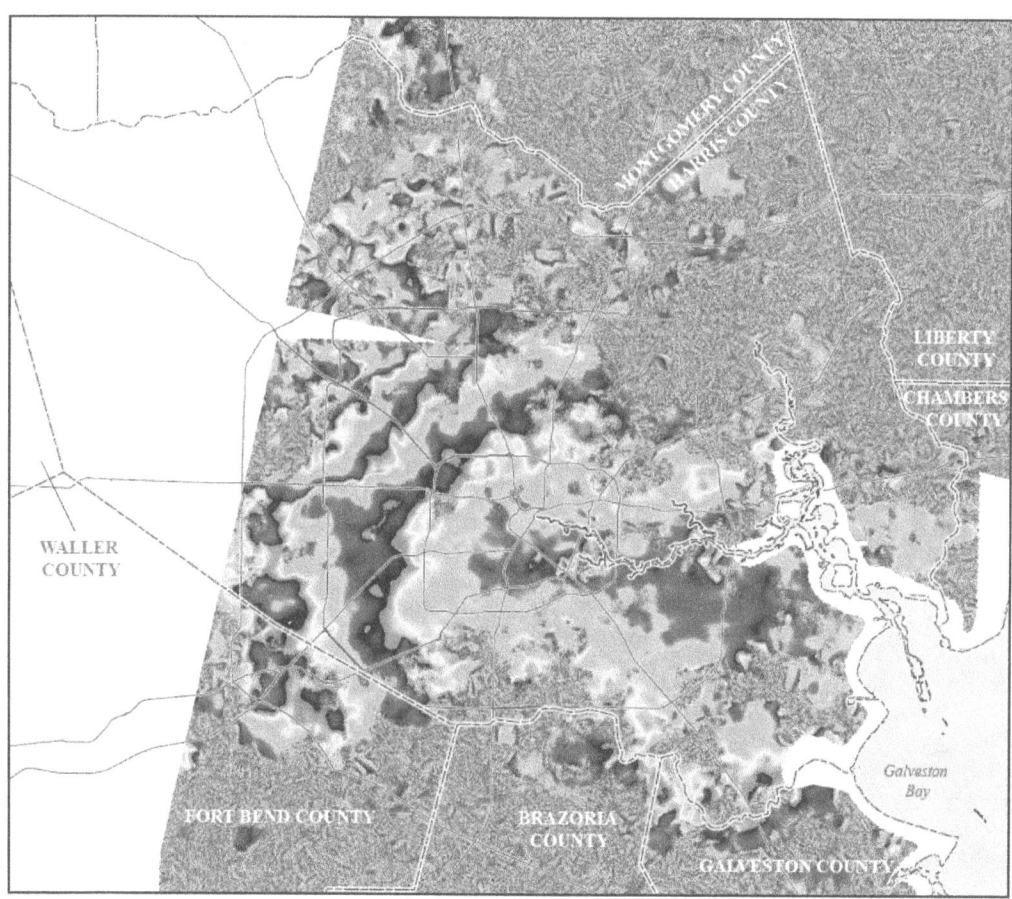

Scientific Investigations Report 2012–5211

U.S. Department of the Interior
U.S. Geological Survey

Investigation of Land Subsidence in the Houston-Galveston Region of Texas By Using the Global Positioning System and Interferometric Synthetic Aperture Radar, 1993–2000

By Gerald W. Bawden, Michaela R. Johnson, Mark C. Kasmarek, Justin Brandt, and Clifton S. Middleton

Prepared in cooperation with the Harris-Galveston Subsidence District

Scientific Investigations Report 2012–5211

U.S. Department of the Interior
U.S. Geological Survey

U.S. Department of the Interior
KEN SALAZAR, Secretary

U.S. Geological Survey
Marcia K. McNutt, Director

U.S. Geological Survey, Reston, Virginia: 2012

Suggested citation:
Bawden, G.W., Johnson, M.R., Kasmarek, M.C., Brandt, Justin, and Middleton, C.S., 2012, Investigation of land subsidence in the Houston-Galveston region of Texas by using the Global Positioning System and interferometric synthetic aperture radar, 1993–2000: U.S. Geological Survey Scientific Investigations Report 2012–5211, 88 p.

Contents

Figures

Tables

Conversion Factors

SI to Inch/Pound

Multiply	By	To obtain
Length		
centimeter (cm)	0.3937	inch (in.)
meter (m)	3.281	foot (ft)
kilometer (km)	0.6214	mile (mi)
Area		
square meter (m^2)	0.0002471	acre
square meter (m^2)	10.76	square foot (ft^2)
square kilometer (km^2)	0.3861	square mile (mi^2)
Flow rate		
cubic meter per second (m^3/s)	35.31	cubic foot per second (ft^3/s)
cubic meter per day (m^3/d)	35.31	cubic foot per day (ft^3/d)
cubic meter per day (m^3/d)	264.2	gallon per day (gal/d)

Vertical coordinate information is referenced to the North American Vertical Datum of 1988 (NAVD 88).

Horizontal coordinate information is referenced to the North American Datum of 1983 (NAD 83).

Investigation of Land Subsidence in the Houston-Galveston Region of Texas By Using the Global Positioning System and Interferometric Synthetic Aperture Radar, 1993–2000

By Gerald W. Bawden,[1] Michaela R. Johnson,[1] Mark C. Kasmarek,[1] Justin Brandt,[1] and Clifton S. Middleton[2]

Abstract

Since the early 1900s, groundwater has been the primary source of municipal, industrial, and agricultural water supplies for the Houston-Galveston region, Texas. The region's combination of hydrogeology and nearly century-long use of groundwater has resulted in one of the largest areas of subsidence in the United States; by 1979, as much as 3 meters (m) of subsidence had occurred, and approximately 8,300 square kilometers of land had subsided more than 0.3 m. The U.S. Geological Survey, in cooperation with the Harris-Galveston Subsidence District, used interferometric synthetic aperture radar (InSAR) data obtained for four overlapping scenes from European remote sensing satellites ERS-1 and ERS-2 to analyze land subsidence in the Houston-Galveston region of Texas. The InSAR data were processed into 27 interferograms that delineate and quantify land-subsidence patterns and magnitudes. Contemporaneous data from the Global Positioning System (GPS) were reprocessed by the National Geodetic Survey and analyzed to support, verify, and provide temporal resolution to the InSAR investigation.

The interferograms show that the area of historical subsidence in downtown Houston along the Houston Ship Channel has stabilized and that recent subsidence occurs farther west and north of Galveston Bay. Three areas of recent subsidence were delineated along a broad arcuate (bow-shaped) feature from Spring, Tex., southwest to Cypress, Tex., and south to Sugar Land, Tex., with subsidence rates ranging from 15 millimeters per year (mm/yr) to greater than 60 mm/yr. Multiyear interferograms near Seabrook, Tex., within the historical subsidence area and nearby Galveston Bay, show several fringes of subsidence (approximately 85 millimeters from January 1996 to December 1997) in the area; however it is difficult to determine the subsidence magnitude near Seabrook because many of the InSAR fringes were truncated or ill-defined. Horizontal and vertical GPS data throughout the area support the InSAR measured subsidence rates and extent. The subsidence rates for a few GPS stations northwest of Houston began to decrease in 2007, which may indicate that subsidence may be decreasing in these areas.

Introduction

The Houston-Galveston region—comprising Harris and Galveston Counties and adjacent parts of Brazoria, Chambers, Fort Bend, Grimes, Liberty, Montgomery, San Jacinto, Walker, and Waller Counties (fig. 1)—is one of the largest areas of subsidence in the United States (Galloway and others, 1999). Most of the subsidence in the Houston-Galveston region (which includes the greater Houston metropolitan area) has occurred as a direct result of groundwater withdrawals for municipal supply, industrial use, and irrigation that depressured and dewatered the major aquifers in the area, thereby causing compaction of the clay layers of the aquifer sediments (Kasmarek and others, 2010; Johnson and others, 2011). Groundwater has historically been the principal source of water for municipal, industrial, and agricultural uses, and groundwater use in the Houston-Galveston region had increased rapidly for many decades to meet the water needs of the rapidly growing population (Seifert and Drabek, 2006). Since the 1990s, surface water has been increasingly used to meet these water needs and reduce reliance on groundwater resources in the Houston-Galveston region (Kasmarek and Robinson, 2004).

Since 1975, the U.S. Geological Survey (USGS), in cooperation with the Harris-Galveston Subsidence District (HGSD), has been monitoring changes in groundwater elevations and measuring cumulative clay compaction by using extensometers at discrete locations in the Houston-Galveston region. Additionally, since the early 1990s, the HGSD has been monitoring land-surface elevation with the National Geodetic Survey (NGS) by periodically conducting

[1]U.S. Geological Survey.

[2]National Geodetic Survey.

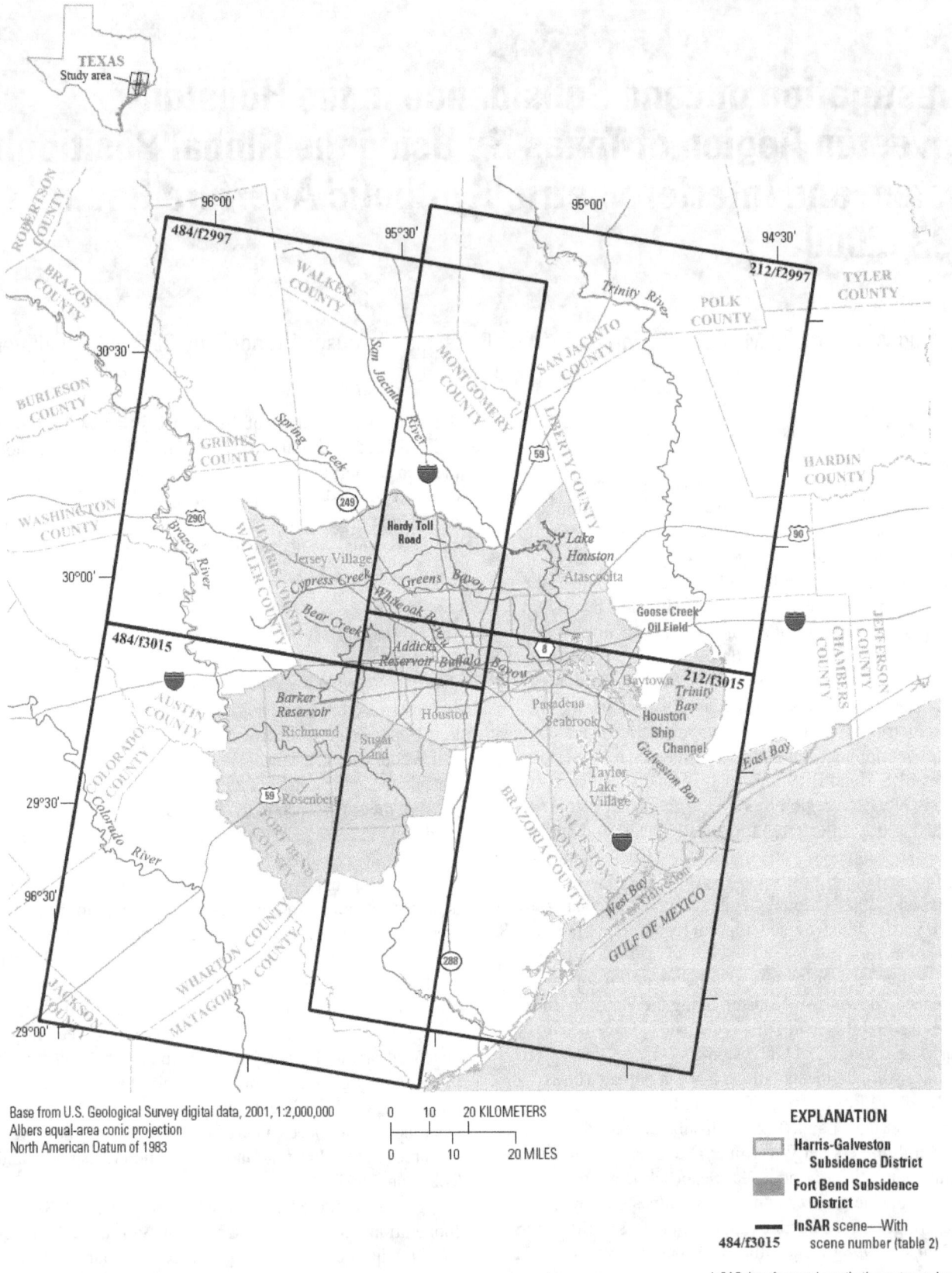

Base from U.S. Geological Survey digital data, 2001, 1:2,000,000
Albers equal-area conic projection
North American Datum of 1983

0   10   20 KILOMETERS

0   10   20 MILES

EXPLANATION

Harris-Galveston
Subsidence District

Fort Bend Subsidence
District

InSAR scene—With
484/f3015   scene number (table 2)

InSAR, Interferometric synthetic aperture radar

Figure 1.   The Houston-Galveston region, Texas, with interferometric synthetic aperture radar (InSAR) scenes, including radar track and frame numbers, and management areas for the Harris-Galveston Subsidence District (2010) and Fort Bend Subsidence District (2009).

first-order leveling surveys and by using land-based Global Positioning System (GPS) methods. But because of the extensive area affected by subsidence (Kasmarek and others, 2010; Johnson and others, 2011), the complexity associated with various magnitudes of groundwater use, and the hydrogeology of the Gulf Coast aquifer, a more robust spatial coverage of deformation was needed to provide information between long-term point measurement locations. The satellite-based interferometric synthetic aperture radar (InSAR) technology is capable of measuring centimeter-level deformation over a large area. Thus, the USGS, in cooperation with the HGSD, used InSAR data and reprocessed regional GPS data to investigate subsidence in the Houston-Galveston region in Texas.

Purpose and Scope

The purpose of this report is to document the magnitude and areal extent of land subsidence patterns in the Houston-Galveston region that were measured by using InSAR data that were obtained from European remote sensing (ERS) satellites ERS-1 and ERS-2 operated by the European Space Agency from July 25, 1992, to December 19, 2000, and reprocessed regional GPS data that were obtained from the NGS for the same time period (European Space Agency, 2012). The historical subsidence in the Houston-Galveston region is mentioned as it relates to the varying methods of determining subsidence within the region. GPS time-series data are discussed to validate the InSAR analysis and provide additional information about the subsidence in regions where InSAR could not provide adequate measurements. This report first discusses the GPS analysis, then the InSAR analysis, and lastly the GPS and InSAR analyses in relation to regional subsidence rates and patterns.

Description of Study Area

The study area exists within four overlapping InSAR scenes of the Houston-Galveston region obtained from two ERS satellites, ERS–1 and ERS–2 (fig. 1). The Houston-Galveston region is approximately 20,000 square kilometers (km^2). Houston, Sugar Land, and Baytown are among the cities in the Houston-Galveston region, which was the sixth largest metropolitan area by population in the United States in 2008 with an estimated 5.7 million inhabitants (United States Census Bureau, 2009b). The population in the greater Houston metropolitan area has grown at a rate exceeding the overall growth rates of Texas and the Nation during 2000–8; population increased by 21.5 percent in the greater Houston metropolitan area compared to 16.7 percent for Texas and 8 percent for the United States during 2000–8 (U.S. Census Bureau, 2009a, b). The population of the greater Houston metropolitan area also increased 19.8 percent during 1990–2000 (United States Census Bureau, 2001), which includes the years during which the subsidence-

related data examined in this report were collected (1992–2000).

The climate in the Houston-Galveston region is subtropical with 30-year normal temperatures ranging from 5 to 34 degrees Celsius and 30-year normal precipitation of 112 centimeters per year (cm/yr) in Galveston and 122 cm/yr in Houston (National Oceanic and Atmospheric Administration, 2008). Land-surface elevation ranges from near sea level upward to about 80 meters (m). The flat land slopes toward the coast decreasing in elevation by about 0.5 m per km and is prone to flooding from heavy rains and tropical storms or from riverine and coastal sources. Subsidence can exacerbate coastal flooding especially along the coast where effects of sea-level rise are compounded by land subsidence (Kreps, 1987). Near the coast of the Houston-Galveston region, eustatic (long term) sea-level (ESL) rise, with estimates of 1 to 1.2 mm/yr, combines with natural (0.05 mm/yr) and human-induced (as much as 2.2 mm/yr) subsidence, thus producing an estimated relative sea-level (RSL) rise in some areas exceeding 2 mm/yr (Paine, 1993).

Hydrogeologic Framework

The Gulf Coast aquifer (also known as the coastal lowlands aquifer system) consists of Miocene and younger unconsolidated sediments in layers of confining and water bearing units (Ryder and Ardis, 2002). The hydrogeologic units composing the Gulf Coast aquifer are thinner and closer to the surface in the northwest, are oriented parallel along the coastline, and dip southeast toward the Gulf of Mexico, increasing in thickness (fig. 2). The hydrogeologic units composing the Gulf Coast aquifer system are the Chicot aquifer, the Evangeline aquifer, the Burkeville confining unit, the Jasper aquifer, the Catahoula Sandstone (confining unit), the Anahuac Formation, and the Frio Formation (Baker, 1979). Water recharging the system as rainfall—primarily in the outcrops in the northwestern part of the system—enters the saturated zone and either travels a short distance, discharging locally to streams and creeks, or moves to deeper zones flowing southeastward to the coast, discharging in wells or into the Gulf of Mexico. The water-bearing units of the Gulf Coast aquifer are composed primarily of laterally and vertically discontinuous fine to coarse-grained sands and gravels with interbedded silts and clays. The Chicot, Evangeline, and Jasper aquifers are the primary sources for municipal groundwater supply in the Houston-Galveston region.

Although the lithology of the Gulf Coast aquifer is primarily unconsolidated, normal faulting does exist as the Gulf of Mexico basin continues to evolve (Chowdury and Turco, 2008). Identifying fault locations and predicting movement of faults in these unconsolidated sediments can be difficult because of the variability and lack of structural control in the underlying rock. The Houston-Galveston region has more than 150 identified active faults, many with rates of movement from 5 to 20 mm/yr (Verbeek and others,

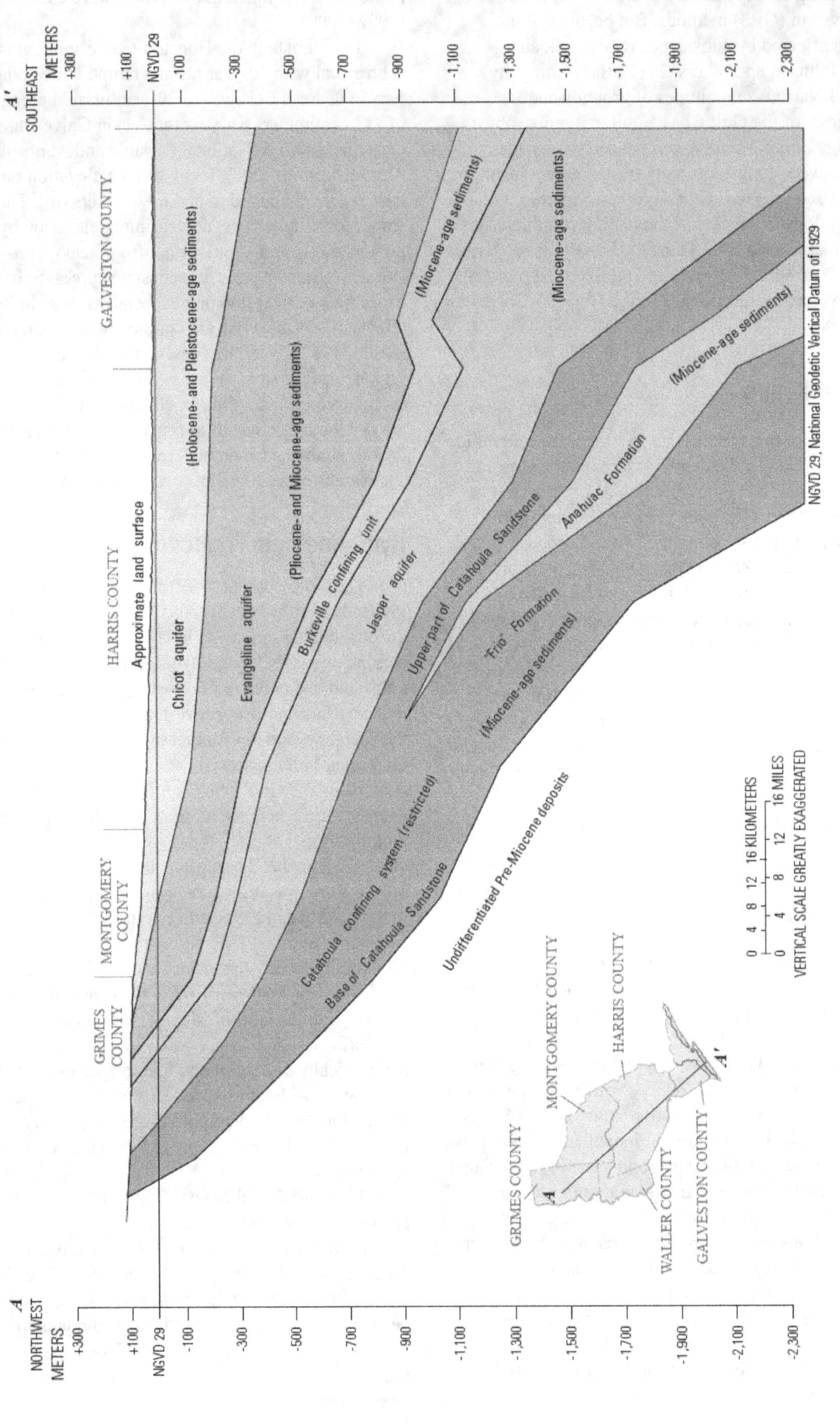

Figure 2.   Hydrogeologic section of the Gulf Coast aquifer system in Harris and adjacent counties, Texas (modified from Baker, 1979, fig. 4).

1979). Although these natural faults have been active in the recent geologic past, accelerated rates of movement were documented in the 20th century (Verbeek and others, 1979). The increased fault movement may be accelerated by fluid withdrawals or be attributable to a period of greater fault activity (Verbeek and others, 1979).

Mechanism and History of Subsidence in the Houston-Galveston Region

Subsidence occurs in the Houston-Galveston region through clay compaction and reflects the pattern of groundwater withdrawals. The weight of the overburden above the aquifer is supported by a combination of pore pressure exerted by the fluid in the aquifer and structural strength provided from the aquifer skeleton, or the granular network of particles in the aquifer (fig. 3). Seasonal groundwater withdrawals result in a scenario where deformation of the aquifer alternates between compression and expansion. When the aquifer is pumped seasonally and aquifer hydraulic heads (measured as groundwater levels in wells) remain above the preconsolidation stress threshold, the aquifer compacts and expands with elastic compaction, and there is no permanent subsidence. When groundwater levels fall below the preconsolidation stress threshold, inelastic compaction of principally the fine-grained deposits (silts and clays) occurs, causing permanent subsidence (Galloway and others, 1999).

The Houston-Galveston region has experienced land subsidence associated with shallow oil and gas extraction and municipal and industrial groundwater withdrawal. Subsidence in the Houston-Galveston region was first correlated with fluid extraction in 1926 at the Goose Creek oil field (Pratt and Johnson, 1926). Although oil and gas withdrawals characteristically result in localized areas of subsidence confined to the field of production, a larger regional area of subsidence is typically associated with shallow groundwater withdrawals. The area of subsidence in the Houston-Galveston region is one of the largest in the United States; by 1979 as much as 3 m of subsidence had occurred and approximately 8,300 km^2 of land had subsided more than 0.3 m (Coplin and Galloway, 1999, p. 40).

Methods used to measure subsidence in the Houston-Galveston region include repeated first-order leveling surveys of existing benchmarks, mechanical measurements of aquifer-system compaction by using borehole extensometers, GPS measurements at NGS Continuously Operating Reference Station (CORS) sites and at portable GPS measuring stations

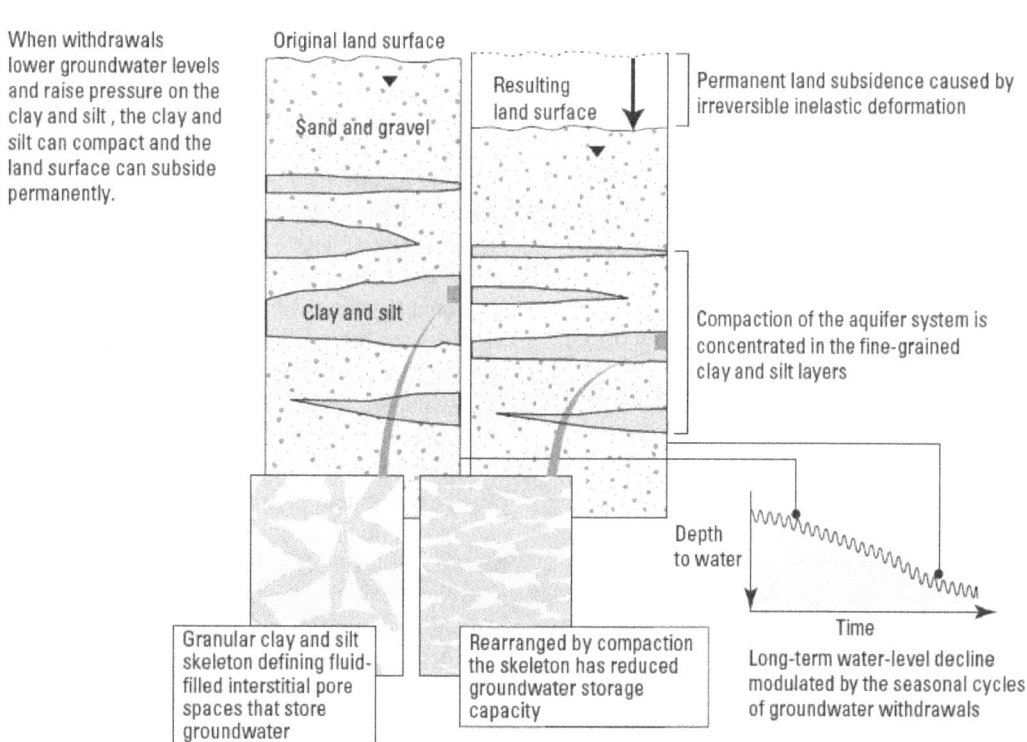

When withdrawals lower groundwater levels and raise pressure on the clay and silt, the clay and silt can compact and the land surface can subside permanently.

Original land surface

Resulting land surface

Sand and gravel

Clay and silt

Permanent land subsidence caused by irreversible inelastic deformation

Compaction of the aquifer system is concentrated in the fine-grained clay and silt layers

Granular clay and silt skeleton defining fluid-filled interstitial pore spaces that store groundwater

Rearranged by compaction the skeleton has reduced groundwater storage capacity

Depth to water

Time

Long-term water-level decline modulated by the seasonal cycles of groundwater withdrawals

Figure 3.   Diagram depicting the mechanism of subsidence in an aquifer composed of vertically discontinuous fine to coarse-grained sands and gravels with interbedded silts and clays (modified from Galloway and others, 1999, p. 9).

called Port-A-Measure (PAM) sites (Harris-Galveston Subsidence District, 2012) (fig. 4), and historical digital elevation model comparison (Kasmarek and others, 2009a). Additionally, groundwater-level change maps have been generated annually since 1977 to document the short-term and long-term effects of groundwater withdrawal on the availability of the groundwater resource. More information on these methods and approaches can be found in Gabrysch and Neighbors (2005), Kasmarek and Houston (2007), Kasmarek and others (2009b), Kasmarek and others (2010), and Johnson and others (2011).

Previous Investigations

InSAR imagery has been used to estimate land-surface subsidence and locate faults in alluvial groundwater basins in the western United States (Galloway and others, 1998; Amelung and others, 1999; Bawden and others, 2001). The spatial and temporal motions associated with both natural and human-induced recharge and discharge can be imaged with well-timed land-surface deformation maps, called interferograms (that is, radar imagery that shows whether the ground has moved closer to or farther from an orbiting satellite). Aquifer structure, extent, and mechanical properties can also be seen or determined from well-timed interferograms (Galloway and Hoffman, 2007). Time-dependent subsidence and uplift from water-level changes have been documented by using InSAR techniques in Los Angeles, California, (Bawden and others, 2001) and in the Santa Clara Valley, Calif. (Ikehara and others, 1998; Galloway and others, 1999; Galloway and others, 2000; Schmidt and Bürgmann, 2003). In the Los Angeles basin, Bawden and others (2001) found that groundwater pumping and artificial recharge produce as much as 60 millimeters (mm) of seasonal uplift and subsidence and discovered that a subsurface barrier to groundwater flow is offset about 1.9 km from the mapped trace of the Newport-Inglewood Fault that bounds the southwest margin of the Santa Anna basin. Other InSAR studies have found previously unrecognized faults as lineaments in the InSAR imagery (Amelung and others, 1999; Catchings and others, 2000; Galloway and others, 2000).

The use of InSAR has also been investigated in the Houston-Galveston region (Stork and Sneed, 2002; Buckley and others, 2003). Stork and Sneed (2002) examined two satellite scenes to develop interferograms for 1996–99 with results matching the patterns of subsidence measured by the extensometers. Buckley and others (2003) developed interferograms for an area in Texas that included the eastern half of Harris County and most of Galveston County and the Galveston Bay area for the period 1996–98. Their study observed areas of subsidence in the western and northwestern parts of Houston as well as near the Seabrook extensometer.

Methodology

InSAR data were acquired, analyzed, and processed to create interferograms following generalized methods described by Schmidt and Bürgman (2003). GPS time-series data were processed and InSAR interferograms were created to assess subsidence patterns in the Houston-Galveston region. These GPS time-series data were used to (1) verify and validate InSAR-derived subsidence patterns and magnitude; (2) compare the GPS-apparent position of benchmark—when subsidence occurs, the position of the benchmark is pulled down and toward nearby subsidence or pushed up and away from uplift; and (3) provide a stable reference frame for the InSAR imagery. InSAR data collection and processing methods for the purpose of measuring land surface subsidence are summarized in appendix 1 (Schmidt and Bürgman, 2003; European Space Agency, 2007).

Global Positioning System Processing

Baseline-Pair Method

In 2010, the NGS uniformly reprocessed all of the regional CORS and PAM GPS station data collected between 1993 and 2009 (Clifton Middleton, National Geodetic Survey, written commun., 2009), and those stations with sufficient data were used for baseline-pair analyses (fig. 5). The approach used to process the GPS data is described by Soler and others (2001) and Soler and Marshall (2002).

The GPS data were provided as baseline pairs for three CORS reference sites. The three CORS reference sites are listed by their GPS station name and four-character GPS site identifier in table 1—Addicks 1795 CORS ARP (ADKS), Lake Houston (LKHU), and Northeast 2250 CORS ARP (NETP). The GPS station names are assigned by the NGS or the Harris-Galveston Subsidence District (National Geodetic Survey, 2011; Harris-Galveston Subsidence District, 2011). The acronym "ARP" is part of GPS station names for some reference sites; the ARP, or antenna reference point, is the physical bottom of the antenna used in the GPS measurement at each site (Landtech Consultants, Inc., 2003). The directional velocities depicting relative movement in millimeters per year for north, east, and up positional changes between a given GPS site and the baseline GPS site, NETP, are listed in table 1. Differences between the north, east, and up positions at the LKHU and NETP reference sites and between those at the ADKS and LKHU reference sites are shown in figs. 6A and 6B, respectively.

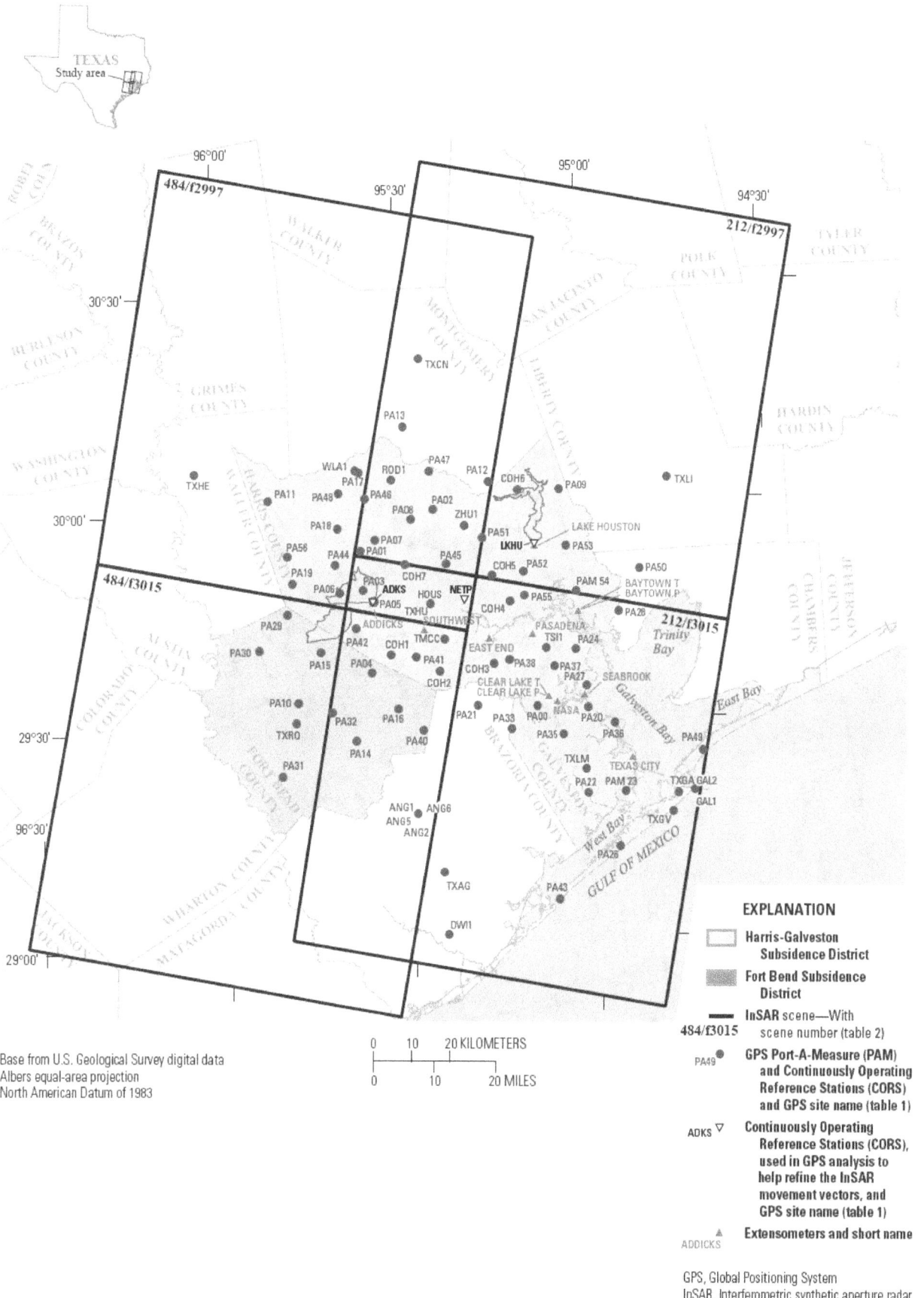

Figure 4. Locations of National Geodetic Survey Continuously Operating Reference Station sites, Harris-Galveston Subsidence District Port-A-Measure sites, and extensometer sites in the Houston-Galveston region, Texas.

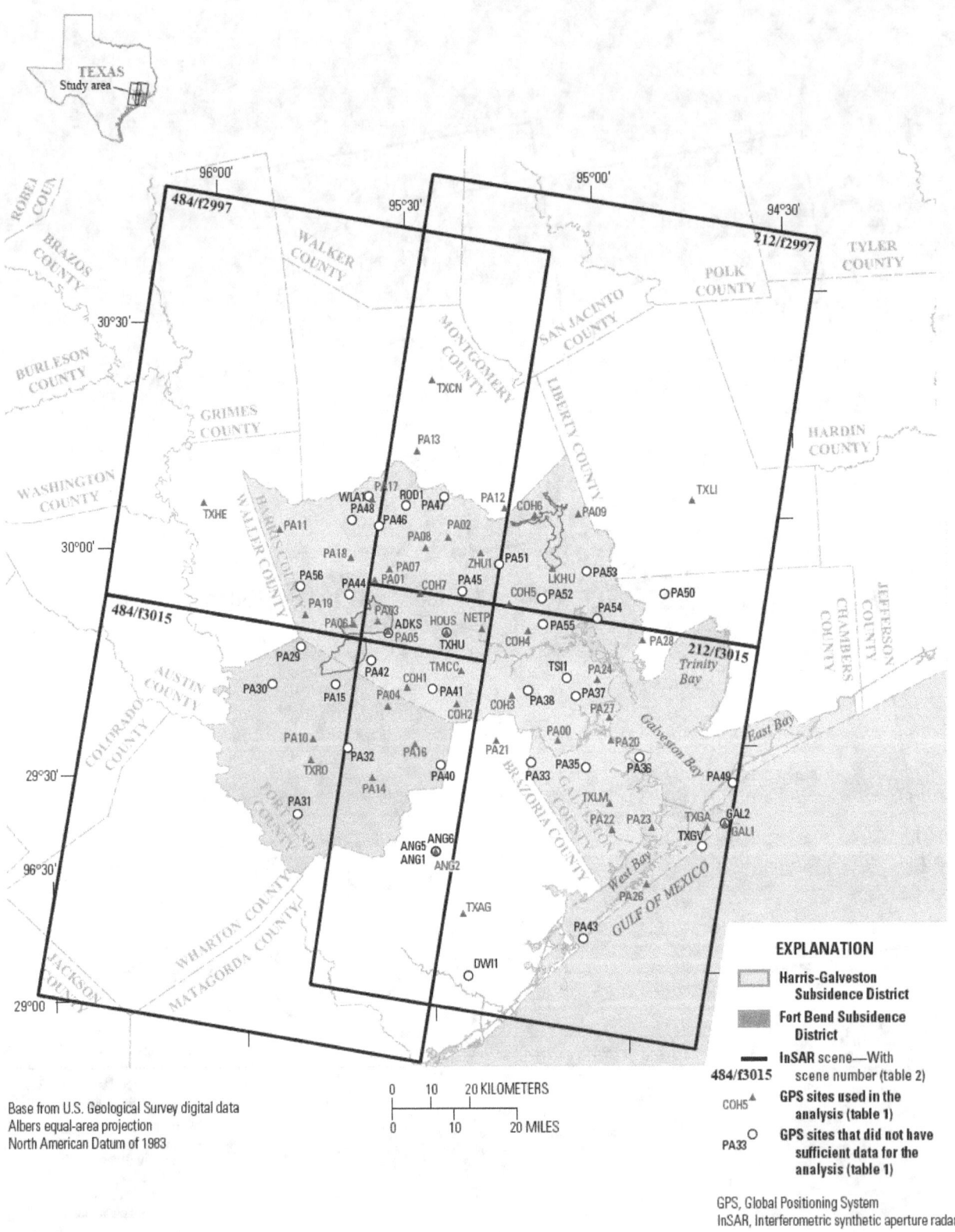

Figure 5.   The National Geodetic Survey (NGS) Continuously Operating Reference Stations (CORS) and Harris-Galveston Subsidence District Port-A-Measure (PAM) network of Global Positioning System (GPS) stations used (pink triangle) and not used (white circle) to evaluate subsidence amounts, Houston-Galveston region, Texas.

Table 1. Global Positioning System (GPS) stations used to validate subsidence magnitudes determined by using interferometric synthetic aperture radar (InSAR) with directional velocities in millimeters per year for north, east, and up positional changes between a given GPS station and the Northeast 2250 Continuously Operating Reference Station antenna reference point (Northeast 2250 CORS ARP).

[Velocity in millimeters per year (mm/yr); NA, not applicable; ----- indicates no velocity determined because the Northeast 2250 CORS ARP station (NETP) was held fixed as the local baseline pair reference frame]

GPS site identifier[2]	GPS station name[2]	Longitude	Latitude	Velocity north (mm/yr)	Velocity east (mm/yr)	Velocity up (mm/yr)
ANG[1]	ANGLETON 1	95° 29' 7.26" W	29° 18' 5.00" N	-5.4	-1.1	4.1
ANG2	ANGLETON 2	95° 29' 5.52" W	29° 18' 5.92" N	-0.1	0.6	-5.3
ANG5	ANGLETON 5	95° 29' 6.00" W	29° 18' 5.00" N	2.2	0.5	0.7
ANG6	ANGLETON 6	95° 29' 5.52" W	29° 18' 5.92" N	1.4	0.5	4.5
COH1	C OF HOUSTON COOP	95° 32' 33.36" W	29° 40' 13.20" N	1.8	1.1	-11.9
COH2	HOUSTON 2 COOP	95° 24' 41.75" W	29° 37' 42.69" N	1.2	-1.2	-4.2
COH3	HOUSTON 3 COOP	95° 15' 51.62" W	29° 38' 32.60" N	1.1	0.2	1.0
COH4	HOUSTON 4 COOP	95° 12' 54.57" W	29° 46' 59.39" N	1.0	0.4	3.8
COH5	HOUSTON 5 COOP	95°16'29.91" W	29°50'40.26" N	1.9	-0.0	-1.5
COH6	HOUSTON 6 COOP	95° 11' 5.27" W	30° 2' 23.04" N	1.2	0.7	-5.8
COH7	HOUSTON 7 COOP	95° 29' 47.76" W	29° 52' 38.13" N	2.8	-0.1	-11.4
DWI1	CLUTE COOP	95°24'13.18" W	29°0'48.98" N	-21.4	32.2	-62.7
GAL1	GALVESTON 1	94°44'12.01" W	29°19'46.99" N	0.9	-0.5	-5.8
GAL2	GALVESTON 2	94°44'12.05" W	29°19'48.25" N	0.1	0.6	-2.4
HOUS	HOUSTON	95°25'58" W	29°46'45" N	1.6	-3.5	-9.8
PA00	PAM 00	95° 9' 8.02" W	29° 32' 19.01" N	-0.6	0.2	-3.1
PA01	PAM 01	95° 36' 59.80" W	29° 54' 42.73" N	-0.8	-1.8	-42.8
PA02	PAM 02	95° 24' 57.09" W	30° 0' 2.34" N	1.7	-1.5	-34.1
PA03	PAM 03	95° 36' 48.15" W	29° 49' 14.91" N	1.7	-2.0	-38.4
PA04	PAM 04	95° 35' 48.68" W	29° 37' 49.40" N	-3.5	-0.8	-19.5
PA05	PAM 05	95° 35' 9.24" W	29° 47' 28.33" N	1.3	0.0	-25.7
PA06	PAM 06	95° 40' 40.00" W	29° 48' 58.92" N	1.2	-1.9	-30.9
PA07	PAM 07	95° 34' 35.91" W	29° 56' 10.65" N	2.7	1.2	-41.6
PA08	PAM 08	95° 28' 34.54" W	29° 58' 46.82" N	3.4	-2.4	-29.1
PA09	PAM 09	95° 4' 17.24" W	30° 2' 17.23" N	2.5	0.1	-3.1
PA10	PAM 10	95° 47' 57.01" W	29° 33' 58.97" N	4.1	2.2	-4.9
PA11	PAM 11	95° 51' 54.78" W	30° 1' 55.77" N	1.4	0.1	-5.2
PA12	PAM 12	95° 15' 47.04" W	30° 3' 34.89" N	2.3	1.1	-9.4
PA13	PAM 13	95° 29' 23.94" W	30° 11' 41.29" N	0.1	0.4	-18.3
PA14	PAM 14	95° 38' 38.75" W	29° 28' 25.14" N	4.3	-3.2	-7.1
PA15	PAM 15	95° 43' 11.11" W	29° 40' 59.66" N	1.4	-0.5	-10.7
PA16	PAM 16	95° 31' 38.05" W	29° 32' 40.05" N	2.7	-8.7	-10.3
PA17	PAM 17	95° 36' 55.03" W	30° 5' 28.17" N	0.2	-0.6	-20.2
PA18	PAM 18	95° 40' 41.59" W	29° 57' 53.75" N	1.5	0.1	-24.7
PA19	PAM 19	95° 48' 19.21" W	29° 50' 28.01" N	0.7	0.1	-10.6
PA20	PAM 20	95° 0' 47.64" W	29° 31' 58.44" N	-0.1	1.7	-0.3
PA21	PAM 21	95° 18' 43.44" W	29° 32' 43.66" N	2.5	-1.8	-5.5
PA22	PAM 22	95° 1' 14.53" W	29° 20' 4.26" N	2.4	-2.0	-5.1
PA23	PAM 23	94° 55' 3.97" W	29° 20' 6.27" N	-2.7	0.1	-1.2
PA24	PAM 24	95° 2' 26.78" W	29° 40' 7.65" N	1.5	0.2	0.7
PA26	PAM 26	94° 56' 17.94" W	29° 12' 37.12" N	1.7	-0.1	-2.2
PA27	PAM 27	95° 0' 55.95" W	29° 34' 59.28" N	-0.3	3.1	-6.3
PA28	PAM 28	94° 55' 3.44" W	29° 45' 4.37" N	-1.6	1.2	0.5
PA29	PAM 29	95° 49' 19.84" W	29° 46' 8.45" N	2.1	-4.8	-13.4
PA30	PAM 30	95° 54' 6.87" W	29° 41' 21.29" N	3.9	-2.5	-6.8

Table 1.   Global Positioning System (GPS) stations used to validate subsidence magnitudes determined by using interferometric synthetic aperture radar (InSAR) with directional velocities in millimeters per year for north, east, and up positional changes between a given GPS station and the Northeast 2250 Continuously Operating Reference Station antenna reference point (Northeast 2250 CORS ARP).—Continued

[Velocity in millimeters per year (mm/yr); NA, not applicable; ----- indicates no velocity determined because the Northeast 2250 CORS ARP station (NETP) was held fixed as the local baseline pair reference frame]

GPS site identifier[2]	GPS station name[2]	Longitude	Latitude	Velocity north (mm/yr)	Velocity east (mm/yr)	Velocity up (mm/yr)
PA31	PAM 31	95° 50' 54.13" W	29° 23' 52.85" N	8.8	-23.6	1.1
PA32	PAM 32	95° 42' 26.27" W	29° 32' 26.15" N	0.9	-1.3	6.2
PA33	PAM 33	95° 13' 24.81" W	29° 29' 23.67" N	-2.0	2.1	7.6
PA35	PAM 35	95° 4' 56.75" W	29° 28' 21.40" N	1.7	1.4	5.9
PA36	PAM 36	94° 56' 29.82" W	29° 29' 39.02" N	1.2	4.8	-0.4
PA37	PAM 37	95° 6' 3.61" W	29° 37' 50.52" N	3.3	-8.7	-2.2
PA38	PAM 38	95° 13' 22.60" W	29° 38' 57.37" N	34.9	-2.4	0.4
PA40	PAM 40	95° 27' 44.96" W	29° 29' 35.84" N	-5.3	0.2	-7.9
PA41	PAM 41	95° 28' 31.77" W	29° 39' 42.85" N	12.0	2.9	-10.2
PA42	PAM 42	95° 38' 7.22" W	29° 43' 56.94" N	-4.5	-11.4	-8.5
PA43	PAM 43	95° 6' 38.11" W	29° 5' 35.67" N	0.3	3.5	-0.3
PA44	PAM 44	95° 41' 12.67" W	29° 52' 48.45" N	7.8	-4.5	-12.5
PA45	PAM 45	95° 23' 7.60" W	29° 52' 33.21" N	2.0	-1.7	5.8
PA46	PAM 46	95° 36' 0.18" W	30° 1' 47.87" N	0.5	-0.1	-22.6
PA47	PAM 47	95° 25' 24.73" W	30° 5' 22.37" N	2.8	-2.8	-12.9
PA48	PAM 48	95° 40' 18.13" W	30° 2' 43.27" N	-1.6	1.8	-11.5
PA49	PAM 49	94° 42' 5.47" W	29° 25' 20.79" N	2.2	2.6	-1.1
PA50	PAM 50	94° 51' 21.72" W	29° 50' 54.02" N	1.1	-2.1	1.4
PA51	PAM 51	95° 17' 3.08" W	29° 55' 57.14" N	4.1	0.2	14.2
PA52	PAM 52	95° 10' 36.22" W	29° 51' 7.27" N	2.3	-0.2	8.5
PA53	PAM 53	95° 3' 26.20" W	29° 54' 28.88" N	3.9	0.5	-0.2
PA54	PAM 54	95° 2' 3.77" W	29° 48' 5.27" N	2.3	-2.3	2.2
PA55	PAM 55	95° 10' 37.89" W	29° 47' 39.06" N	1.0	3.3	6.5
PA56	PAM 56	95° 49' 0.34" W	29° 54' 9.40" N	0.4	4.5	-7.0
ROD1	SPRING	95° 31' 36.00" W	30° 4' 19.99" N	3.7	-0.1	-18.0
TXAG	ANGLETON	95° 25' 7.97" W	29° 9' 49.97" N	3.3	-0.1	-1.8
TMCC	----	95° 23' 42.82" W	29° 42' 8.42" N	0.9	-0.2	-1.5
TSI1	DEERPARK COOP	95°7'21.25" W	29°40'23.48" N	0.9	1.6	2.3
TXCN	CONROE	95° 26' 28.00" W	30° 20' 56.00" N	1.3	0.7	-11.7
TXGA	GALVESTON	94° 46' 21.00" W	29° 19' 39.97" N	0.1	2.2	-5.1
TXGV	GALVESTON 877 1510	94° 47' 21.45" W	29° 17' 6.48" N	3.0	-1.9	7.2
TXHE	HEMPSTEAD	96° 3' 47.99" W	30° 5' 56.00" N	1.0	2.3	-4.1
TXHU	HOUSTON RRP2	95°25'57.90" W	29°46'45.19" N	0.7	-0.4	-3.4
TXLI	LIBERTY	94° 46' 14.99" W	30° 3' 20.99" N	0.7	1.7	-2.9
TXLM	LA MARQUE	95° 01' 25.26" W	29° 23' 32.00" N	0.4	0.5	-5.5
TXRO	ROSENBERG	95° 48' 25.99" W	29° 31' 8.00" N	0.6	2.8	-6.7
WLA1	TOMBALL COOP	95°37'30.60" W	30°5'47.09" N	-3.2	3.3	-17.8
ZHU1	HOUSTON WAAS 1	95° 19' 53.10" W	29° 57' 42.81" N	2.9	-0.8	-7.9
ADKS	Addicks 1795 CORS ARP	95° 35' 11.04" W	29° 47' 27.47" N	-1.5	2.5	-0.2
LKHU	Lake Houston	95° 8' 44.69" W	29° 54' 48.44" N	-2.7	0.1	0.7
NETP	Northeast 2250 CORS ARP station	95° 20' 3.17" W	29° 47' 28.14" N	-----	-----	-----

[1]Italicized font indicates the site was not used for the InSAR analysis.

[2]Station names and site identifiers are from National Geodetic Survey CORS site search - http://www.ngs.noaa.gov/CORS/ or Harris-Galveston Subsidence District - Subsidence Charts - http://mapper.subsidence.org/Chartindex.htm.

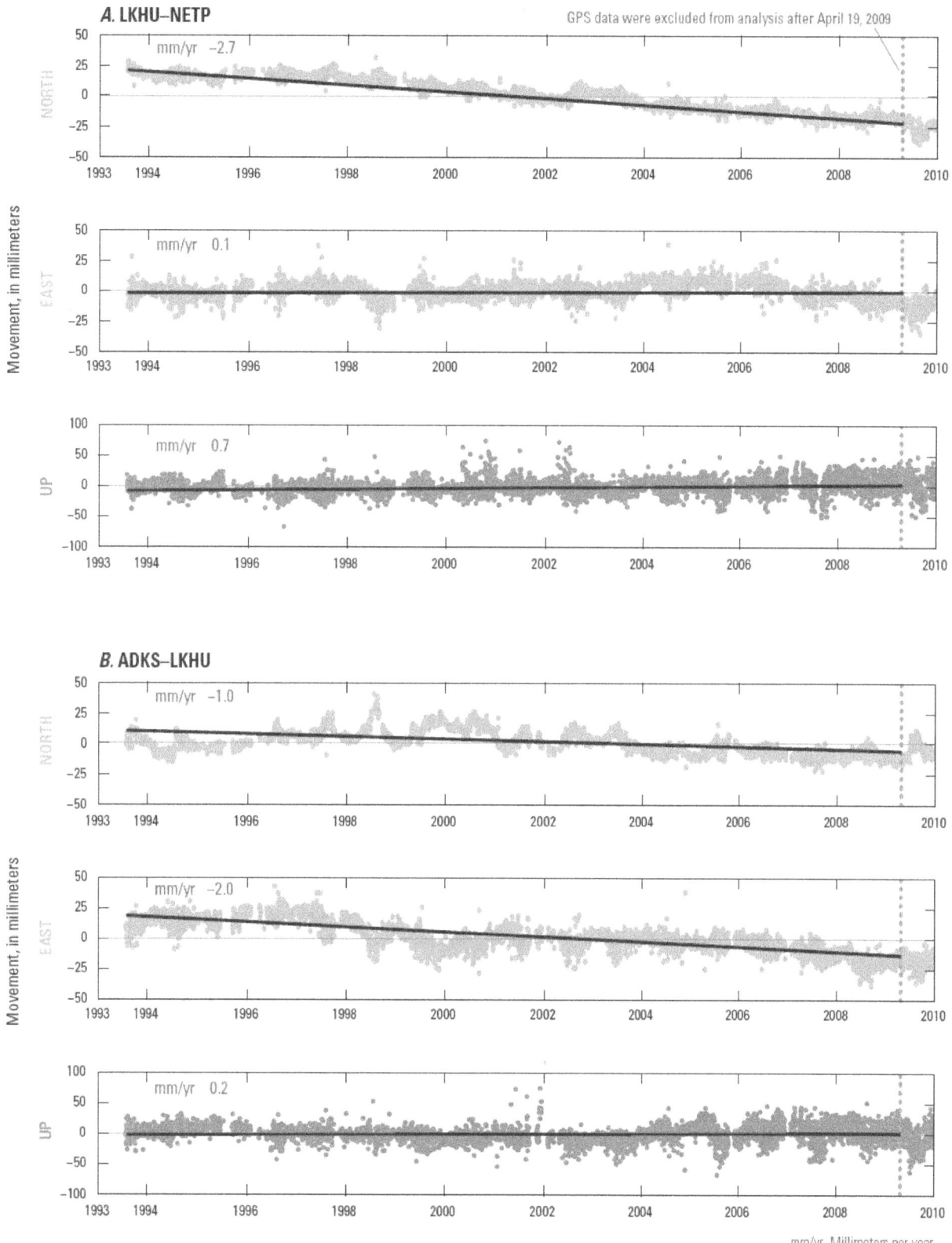

Figure 6.   Global Positioning System (GPS) time series of relative movement with least-squares linear-regression lines depicting the directional movement in millimeters per year (mm/yr) between A, Continuously Operating Reference Stations (CORS) Lake Houston (LKHU) and Northeast 2250 CORS ARP (NETP) and B, Addicks 1795 CORS ARP (ADKS) and Lake Houston (LKHU).

Each pair of fixed GPS stations forms a baseline and was analyzed for relative positional change over time in the north, east, and up components measured to subcentimeter precision. Baseline pairs measured over time for LKHU–NETP and ADKS–LKHU CORS sites are represented as a GPS time series (figs. 6A and 6B, respectively). When there is no change in the relative position between paired stations, the apparent velocity is zero. If there was an observed change in position associated with subsidence or uplift, however, then a GPS station moving northeast or away from the reference station because of subsidence, for example, would show an increased distance in both the north and east components and a decreased distance in the up component as the GPS station is pulled in towards a subsidence feature (vertical motion associated with aquifer compaction) (fig. 7).

To minimize the influence of seasonal fluctuations on the long-term velocity pattern, GPS-measured velocities were determined only for time series with more than 3.5 years of data (Blewitt and others, 1995; Bawden and others, 2001). Concurrent baseline pairs that differed by more than 50 mm from the time series were removed, and a least-squares linear-regression equation (Helsel and Hirsch, 2002) was computed for the remaining data for each of the directional components (north, east, and up) of the data to compute the horizontal and vertical velocities at each station. Time series that had appreciable steps (an appreciable change (20 mm or larger) in velocity from one velocity value to the next) in velocity in either the horizontal or vertical component were excluded from this analysis. For example, data were excluded when it was unclear if the positional change was a result of land deformation or error. Examples of GPS error sources include multipath noise (GPS signal reflected and diffracted off surfaces with multiple signals reaching the antenna), antenna calibration errors, atmospheric effects, and GPS hardware issues (Byun and others, 2002). An example of the unclear source of positional change can be seen in the time series for GPS sites ANG1 and NETP (fig. 8); the north component shifted about 20 mm to the south during 2004–8; neither an east nor up component of movement was evident. The northerly velocity component for ANG1 is about -5 mm/yr and was affected by this shift, which might have been an artifact given the lack of change in either the east or up components. If the questionable data were removed from the velocity calculation, the velocity would be close to zero. Site ANG1 was not used in the analysis because the time series length did not meet the minimum requirement of 3.5 years, but it underscores how numerical artifacts and short time series can introduce errors into the site velocities.

Selection of a Stable Reference Frame

Defining a stable reference frame for the GPS data is essential to understand how changes in the GPS station positions relate to motion of the crust of the Earth resulting from hydrologic and tectonic processes. There are three basic categories of reference frames: absolute, relative, and baseline pair (Rothacher, 2001).

An absolute reference frame integrates GPS data from the nationwide NGS CORS network in order to understand how a local geodetic network responds on a continental scale. For this report, an absolute reference frame allows the user to understand how the Houston-Galveston region is moving (subsiding) with respect to a presumed stable North American continent. The tradeoff for using an absolute reference frame is that there needs to be an established GPS network with stations that are within a few hundred kilometers of the target networks, such that a similar suite of GPS satellites are simultaneously observed by both the reference and study area GPS receivers. The number of GPS stations and their distances from the target networks factor into the positional error in an absolute reference frame (Snay and others, 2002).

Similarly, a relative reference frame network can be established by holding the positions fixed at a number of GPS stations outside of the deformation area such that positional changes can be measured among the stations within the local network. For the data analyzed in this report, a relative network of GPS stations outside of the deforming regions would have had to have been established in 1993, which was not the case.

The third baseline-pair approach calculates the positional changes between a well-established GPS reference station (present prior to and throughout the time period examined and has a period of record greater than 3.5 years) within the network and each subsequent GPS station. A network can have more than one GPS reference station, which is the station held fixed in the baseline pair analysis. For any baseline pair of stations that are collecting data at the same time, it is possible to calculate very accurate (subcentimeter) distances between the GPS reference station and each of the CORS and PAM stations. For this analysis, a baseline-pair reference frame enabled the measurement of subcentimeter positional changes between the original three GPS stations established in 1993 (ADKS, LKHU, and NETP) and all subsequent regional GPS stations, including the PAM stations.

Interferometric Synthetic Aperture Radar Processing

Interferogram Overview

For this report, InSAR data were acquired, analyzed, and processed to create interferograms. The processing of satellite synthetic aperture radar (SAR) scenes to generate interferograms is a multistep process that transforms the initial radar echoes emitted by the satellite into georeferenced radar images that can detect subcentimeter changes in the land-surface position. The first step focuses two SAR scenes that imaged the same area on the ground from the same viewpoint.

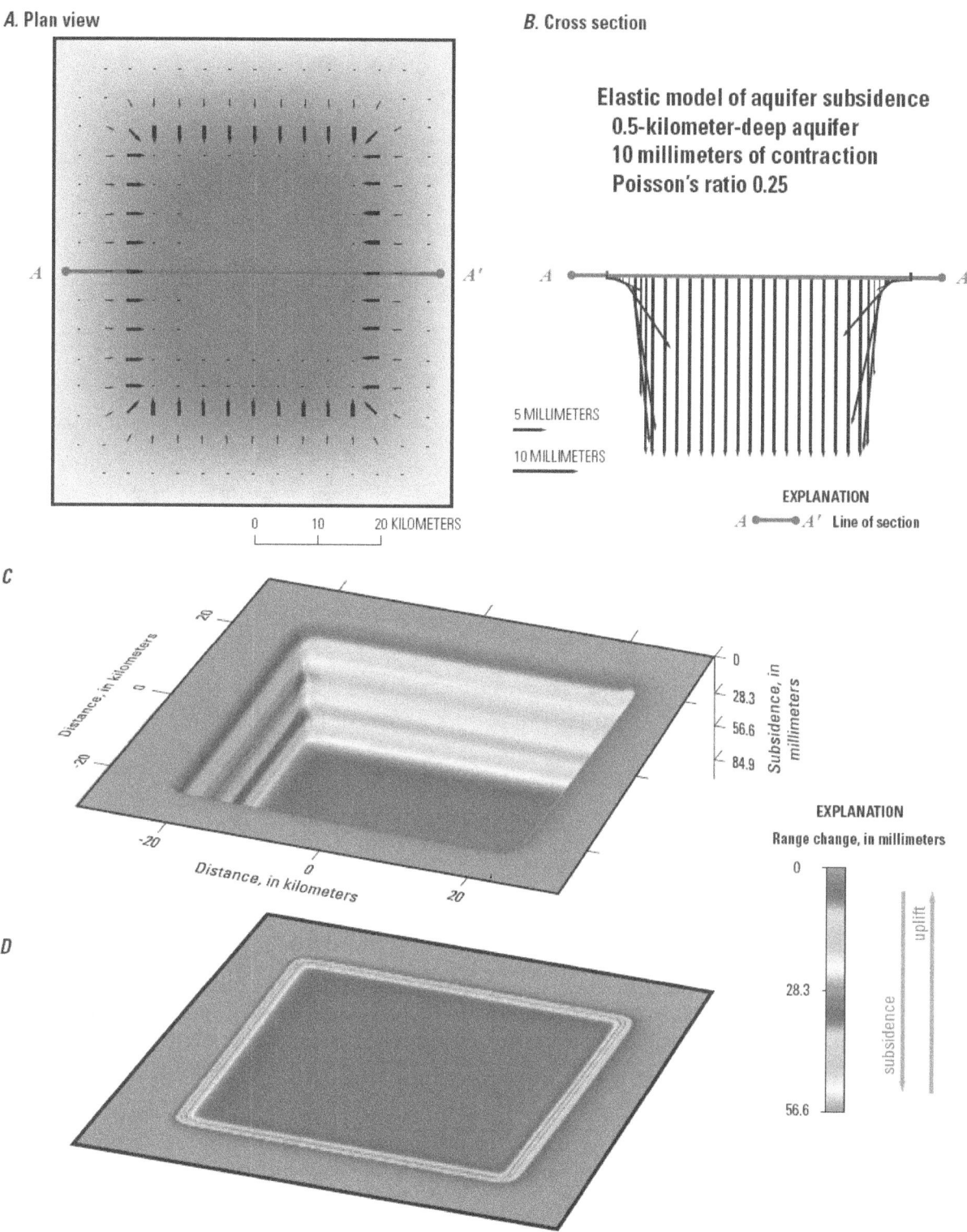

Figure 7. Conceptual depiction of horizontal and vertical movement associated with aquifer system compaction. *A*, Map view of a simulated aquifer subsidence for 10 millimeters (mm) of downward vertical movement, with arrows depicting horizontal movement in and towards the area with the maximum subsidence corresponding to profile *A–A'*. *B*, Cross-section showing the lateral and vertical displacements for profile *A–A'*. *C*, Simulated interferometric synthetic aperture radar (InSAR) image of the modeled aquifer system compaction with vertical exaggeration to visualize the number of individual fringes on the margins of the basin. *D*, Simulated InSAR image of the modeled aquifer system compaction where the InSAR fringe gradient is the largest on the margins of the basin.

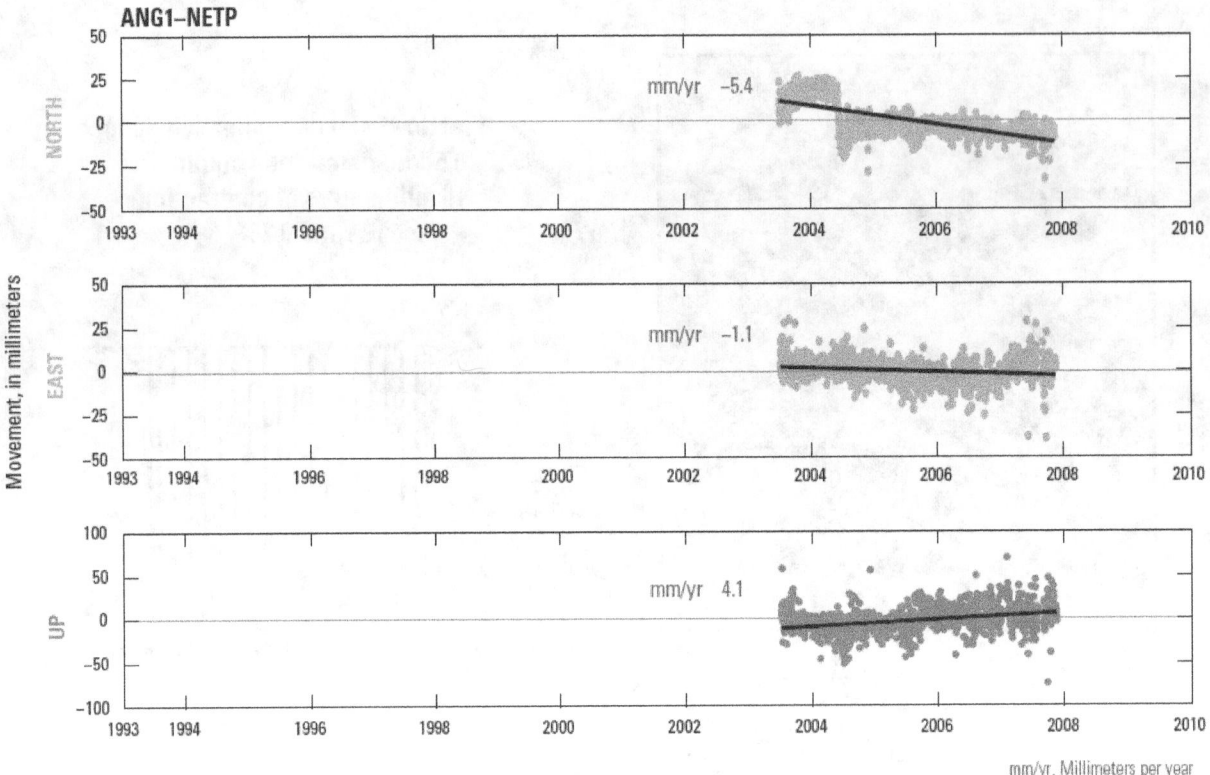

Figure 8.   Global Positioning System (GPS) time series of relative movement with least-squares linear-regression lines depicting the directional movement in millimeters per year (mm/yr) between GPS station ANG1 and Continuously Operating Reference Station Northeast 2250 CORS ARP (NETP).

The two SAR scenes for this report were acquired from different orbits from satellites ERS-1 and ERS-2. The orbital period, or the time it takes for the satellite to return to the same point in space, was 35 days. Next, the two SAR scenes are precisely aligned to each other and the scenes are differenced, or values for the same pixel location are subtracted between the scenes, to create the interferogram (European Space Agency, 2007).

An interferogram shows the measured range change (distance change) between the satellite antenna and any stable point on the ground such as a building or rock outcrop. The satellite antenna emits a sinusoidal wave signal that travels from the satellite to the ground. The wave signal interacts with and bounces off of the land surface and sends a fraction of the signal back to the satellite. The satellite records the length of time for the signal to travel to the land surface and return to the satellite, how much signal returned (amplitude), and what fraction of one full sine-wave cycle returned (phase). The amplitude is not used in calculating the range change. The interferogram maps the change in phase distance for each pixel on the ground and is typically visualized through

repeating color bands, or fringes, which represent changes in phase distance (European Space Agency, 2007) (fig. 9). Only interferograms with good coherency (indicated by an absence of appreciable signal scatter between the original wave signal and the signal returned from the land surface to the satellite) are useful; for these interferograms, image focusing is able to combine the individual radar echoes into one coherent image (appendix 1).

In this report, each color band of the C-band sensor represents 28.3 mm of either uplift or subsidence depending on the color-band pattern (unless otherwise noted, one complete color cycle (fringe) on an interferogram is half the wavelength (56.6 mm) of the C-band sensor, or 28.3 mm and the resolution of one pixel is 30 m^2). Relative displacement of land surface between regions on the map can be measured by counting the number of full (one complete color cycle) and partial color fringes between the two regions of interest on the interferogram and then multiplying by the fringe-scale factor (28.3 mm). For example, an interferogram depicting three repeating fringes of the same subsidence color-band pattern would represent a total of about 85 mm of subsidence. Ideally,

A. Simplified interferogram showing uplift

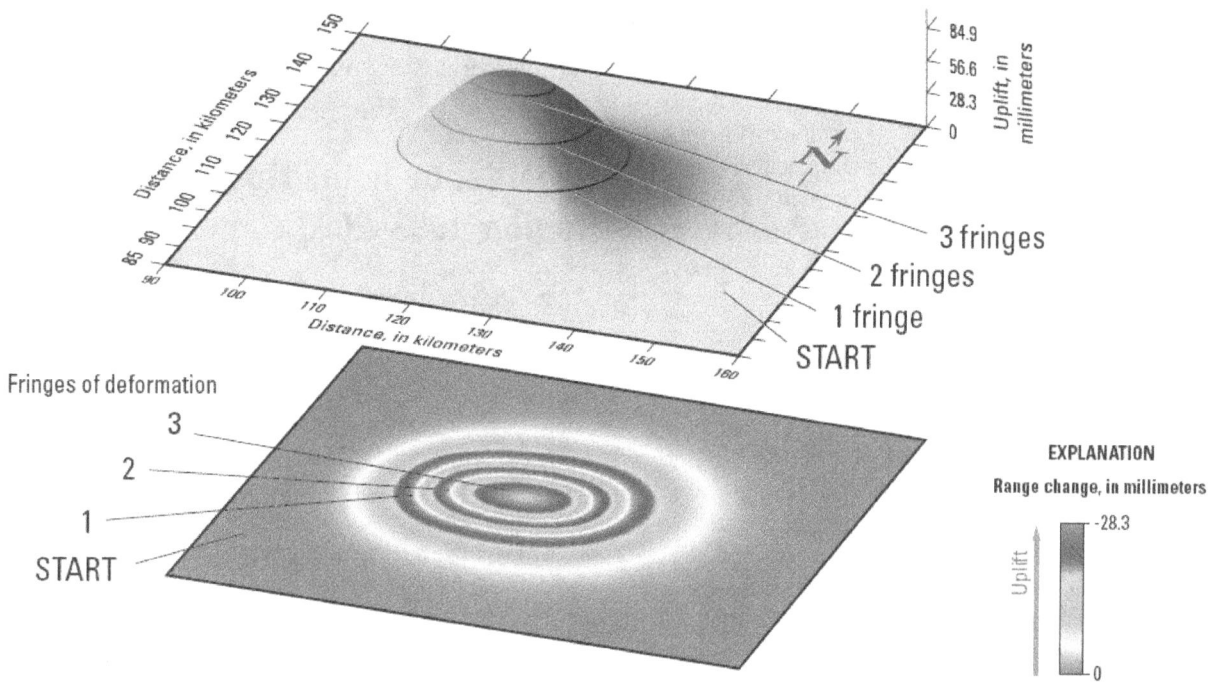

Fringes of deformation

EXPLANATION

Range change, in millimeters

B. Steps to interpret an interferogram

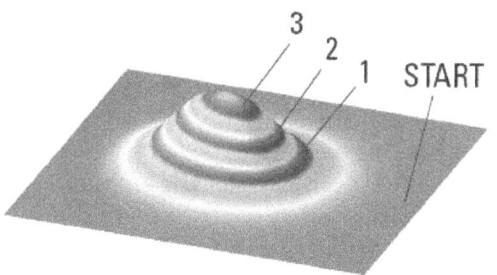

- Count the number of fringes

  From the outer red area ("START") to the
  center, three red fringes can be counted.

- Multiply the number of fringes by 28.3 millimeters to estimate deformation

  3 x 28.3 = 84.9 millimeters

- Determine range change direction from the scale bar

  The color sequence from the outer edge inward
  is red-orange-yellow-green-blue-violet, which
  corresponds to uplift and a decreased distance
  between the ground and the satellite (range).

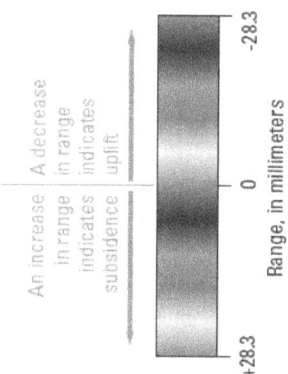

Figure 9.   Simplified interferogram. *A,* Approximately three fringes of deformation equalling 84.9 millimeters (mm) of uplift. *B,* Steps to interpret an interferogram.

if no land surface deformation occurred between the two radar scenes and there was a clear sky, then the interferogram will be a uniform color without any partial fringes except for the influence of local noise sources, such as vegetation, atmospheric moisture, and topography (see appendix 1 for an in-depth discussion on noise sources). If there were surface deformation sometime between when the two radar scenes were acquired, however, then the resulting fringes on the interferogram map would change between neighboring pixels on the ground. Appendix 1, "Interferometric Synthetic Aperture Radar Processing and Error Analysis," provides a more comprehensive discussion on the InSAR processing approach used in this report.

Interferometric Synthetic Aperture Radar Error Considerations

Techniques used to create interferograms from satellite SAR data have limitations with regards to measuring surface motion in regions that are moderately to densely covered in vegetation because the presence of dense vegetation, the growth of vegetation, and other changes in vegetation can appear to alter the phase distance. Atmospheric moisture (clouds), high humidity, and topography can also introduce signal noise (Mateus and others, 2011), but filtering algorithms can enhance the signal quality used to develop interferograms and minimize noise associated with vegetation. Many of these typical noise (error) sources are found in the Houston-Galveston region. To mitigate the effects of changes in vegetation, interferograms were developed from radar images acquired close in time. The Houston-Galveston region SAR scenes were selected to minimize the time period between each acquisition and were selected to study the land surface during the same time of the year, that is, during the fall and winter when there are fewer leaves, which allows the radar signal to reach the ground.

The largest single source of data noise, or error, for the Houston-Galveston region was atmospheric moisture associated with clouds. In general, atmospheric moisture delays the radar signal as it travels from the satellite to the ground and back. This delay increases the radar signal travel time and the calculated range distance and can produce artifacts in the interferograms that equate to as much as 300 mm of delay for the most extreme conditions. Figure 10 includes an interferogram with few clouds and minimal noise (fig. 10A) and an interferogram with storm clouds over part of the Houston-Galveston region (fig. 10B). The error associated with atmospheric moisture was mitigated by combining or stacking radar scenes to average out any atmospheric noise and by avoiding SAR scenes from days with measureable precipitation. Interferograms with severe atmospheric noise were not used in the analysis. High humidity delays the radar signal in a manner similar to atmospheric moisture, but its magnitude is smaller. To reduce error from humidity, the same techniques used to account for atmospheric moisture were applied (European Space Agency, 2007).

Land-surface topography is examined during the error-assessment process. To minimize topographic errors, combinations of SAR scenes with the least orbital variations were used. Because the Houston-Galveston region is flat, the topography errors were negligible (Schmidt and Bürgman, 2003; European Space Agency, 2007).

Subsidence in the Houston-Galveston Region, 1993–2000

NGS GPS data were used to verify subsidence during 1993–2000 in the Houston-Galveston region. The NGS reprocessed all available GPS data (CORS and PAM stations) from January 1, 1993, through December 31, 2009, by using the baseline-pair approach. NGS data from January 1, 2001, through April 19, 2009, were included because these additional years of data collection provided a larger dataset for comparison purposes. Land-surface horizontal and vertical vectors were generated from the time series velocity components from the baseline-pair analyses by examining each station with respect to the NETP CORS reference station. Patterns in the horizontal and vertical GPS time series exhibited general agreement with the InSAR-measured subsidence.

Interferograms were constructed from the SAR scenes controlled by scene availability and quality. The InSAR-derived magnitudes and extents of subsidence were determined over multiple time intervals and verified by contemporaneous GPS time-series analyses. InSAR interferograms showed that the part of the Houston-Galveston region with the maximum amount of historical subsidence during 1913–17 to 2001 of approximately 2 m or more (central to southeastern Harris County, extending south toward Galveston Bay) (Kasmarek and others, 2009a) is now relatively stable. The relative GPS vertical velocities in this area also show little vertical motion relative to the NETP reference station, corroborating the InSAR findings.

Global Positioning System Analysis

Global Positioning System Baseline-Pair Analyses

The NGS reprocessed all of the GPS data for the region, including those from the CORS and PAM stations, with respect to the original three CORS reference stations: ADKS, LKHU, and NETP. Baseline differences in north, east, and up between paired stations LKHU and NETP, and stations ADKS and LKHU, were produced for every epoch date (time associated with the coordinates of the base stations) between 1993 and 2010. Baseline differences for each epoch were also determined as new CORS and PAM stations were added to the

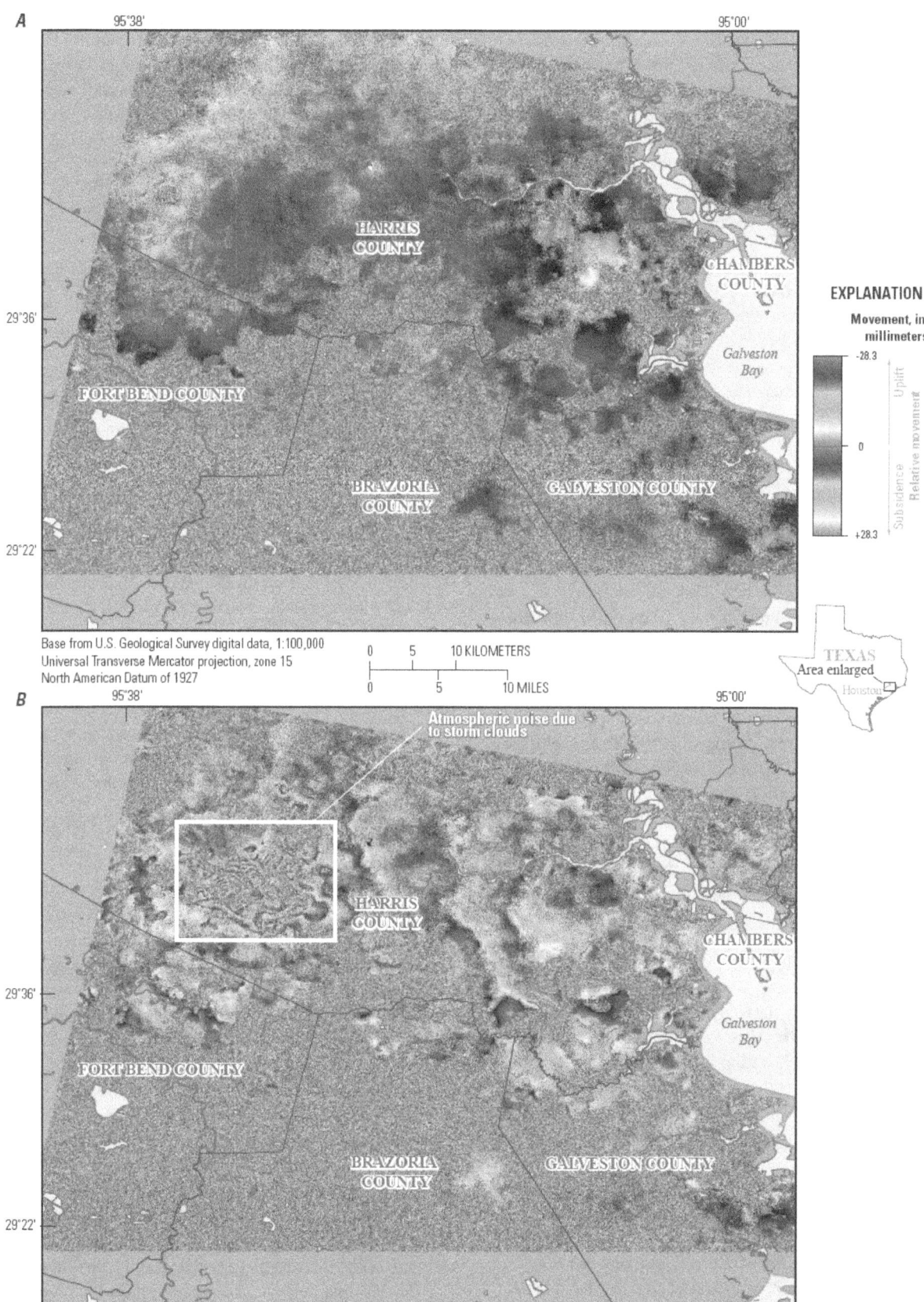

Figure 10. Examples of interferograms. *A*, Few clouds were present and there was minimal noise, Scene T212-F3015 for January 14, 1997, to December 30, 1997. *B*, Storm was present (box), Scene T212-F3015 for January 14, 1997, to July 13, 1999. The fringes in this interferogram are the summation of land surface motion (subsidence), noise related to the storm (box), and possible other clouds in the scene.

network of GPS stations. All GPS data were excluded from the analysis after April 19, 2009, because there was a prominent anomaly that was seen in many of the time-series pairs that could not be easily explained but may have been caused by unexplained motion at one or more of the three CORS stations (note the motion in figs. 6A and 6B after the dashed line at April 19, 2009). Given that April 19, 2009, is well beyond the InSAR time intervals, the exclusion of this data does not affect the interpretation of the interferograms presented in this report.

Relative motion between paired stations ADKS and LKHU (fig. 6B) and stations NETP and LKHU (fig. 6A) between 1993 and 2009 was determined, but baselines between ADKS and NETP (fig. 11) were not processed on a regular basis, so it is not possible to fully assess the stability of the three stations together. The LKHU–NETP baseline pair showed that LKHU moved southward towards NETP at rate of 2.7 mm/yr, or more than 40 mm from 1993 to 2009; however, there was no discernible eastward or vertical seasonal motion during this period (fig. 6A).

The ADKS–LKHU baseline pair showed that ADKS moved to the south and away from LKHU at a rate of about 1 mm/yr and to the west at 2 mm/yr with no substantial uplift during 1993–2009 (fig. 6B). This movement equates to about 20 mm of southward motion and 40 mm of westward movement during the time period. The ADKS–LKHU time series showed higher seasonal variability than the LKHU–NETP time series (figs. 6A and 6B), with some seasonal variability in motion approaching 40 mm. Based on the time-series stability and InSAR analysis, NETP was selected as the reference station in the local baseline-pair reference frame, and all regional motion was tracked with respect to this station. The GPS time series analyzed for all GPS stations are available in appendix 2 of this report, which also contains 83 time series that show the baseline-pair time series between NETP and all of the regional stations.

Time-Series Patterns

The GPS time series showed seasonal motion, long-term patterns, and deviations from the long-term horizontal and vertical patterns. The GPS station LKHU exhibited the most notable seasonal motion (most distinguishable primarily in the up component of movement), which oscillates within the 1–3 mm range, in the time series (fig. 6A). Because the effect of seasonal motion on long-term motion patterns diminishes when there are more than 3.5 years of data (Blewitt and others, 1995; Bawden and others, 2001), the GPS station velocities determined for LKHU and several of the other GPS stations are considered robust. Many of the GPS stations had long and stable time series; most of the PAM stations, from PA01 to PA28, had assessable horizontal and vertical patterns of movement. For example, PA01 had a long-term subsidence

rate of about 43 mm/yr with only a minor southerly motion (table 1). PA04 was subsiding at a rate of about 19 mm/yr with a southerly velocity of about 3 mm/yr (table 1). LKHU had no discernible vertical and easterly motion but had a steady southerly velocity of about 3 mm/yr, or more than 40 mm since 1993 (fig. 6A).

Global Positioning System Vertical and Horizontal Velocities

The GPS stations with the greatest subsidence rates from this analysis were clustered to the northwest and west of the Houston-Galveston region (farthest from downtown Houston, which is the part of this region with the most historical subsidence). Stations recording the most horizontal motion, for example, PA08, PA10, PA14, or PA16, were near the margins of the subsidence (fig. 12). Subsidence rates measured at GPS stations PA01, PA02, PA03, PA06, and PA07 were in excess of 30 mm/yr (fig. 12). These stations delineated the area of greatest subsidence, which was between the cities of Jersey Village, Tex., and Cypress, Tex., and extended to the south toward Sugar Land, Tex. (PA01, 42.8 mm/yr; PA02, 34.1 mm/yr; PA03, 38.4 mm/yr; PA06, 30.9 mm/yr; and PA07, 41.6 mm/yr). The horizontal velocities north of NETP showed a prominent northerly velocity where stations were pulled towards the regions with greater subsidence. Five GPS stations west of Houston (PA10, PA14, PA16, PA04, and TXRO) had a radial velocity inward (fig. 12); these stations correspond to a subsidence trough. Minimal subsidence and horizontal deformation rates were generally measured at sites near Houston (within the Interstate 610 or inner loop and east toward Galveston Bay) (fig. 12).

InSAR Magnitude and Extent

InSAR imagery was processed and analyzed to measure the surface displacement patterns and magnitudes in the Houston-Galveston region. In total, 61 ERS-1 and ERS-2 SAR scenes on four overlapping scenes (fig. 1) were acquired and processed into 214 interferograms for this study with 27 interferograms showing good coherency from September 2, 1993, to December 3, 2000 (table 2). The SAR acquisition dates ranged from July 25, 1992, to December 19, 2000, and the time spans of the interferograms ranged from 1 (tandem ERS-1/ERS-2 interferograms) to 3,069 days (approximately 8 years). The interferograms described in this report were constructed from data obtained from precise satellite orbits by using the two-pass method (Borik and Novotny, 2011) with a USGS 30-m digital elevation model and GAMMA InSAR software (GAMMA Remote Sensing and Consulting, 2012). The InSAR time period examined overlaps the more recent end of the time periods of previous studies examining surface displacement, or subsidence, in the Houston-Galveston region.

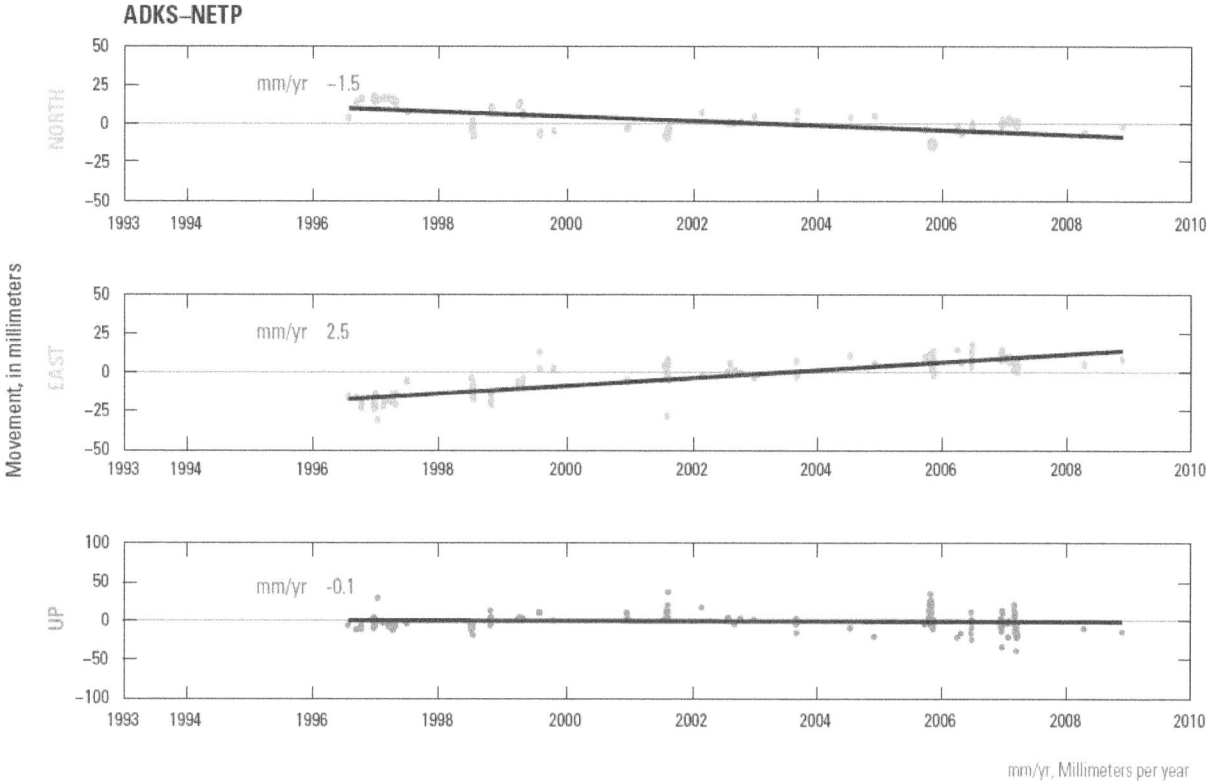

Figure 11.   Global Positioning System (GPS) time series for Continuously Operating Reference Stations (CORS) Addicks 1795 CORS ARP (ADKS) and Northeast 2250 CORS ARP (NETP) with least-squares linear-regression lines depicting directional velocity.

The recent subsidence patterns interpreted for this report are different compared to historical (1906–2000; 1913–17 to 2001) patterns of subsidence published previously (Gabrysch and Neighbors, 2005; Kasmarek and others, 2009a). The historical maximum area of subsidence exists near downtown Houston east along the Houston Ship Channel in southeastern Harris County near Galveston Bay, whereas the InSAR analysis shows that most of the recent subsidence in the area is farther north (along the Harris–Montgomery County line), in western Harris County, and southwest along the Harris–Fort Bend County line (Kasmarek and others, 2009a). Water-level change contours for the Evangeline aquifer between 1990 and 2003 showed a rise near downtown Houston and a decline to the northwest and southwest of Houston by as much as 30 m in a pattern that is similar to the InSAR imagery (fig. 13) but not identical, based on differences between the time periods analyzed by the datasets (Kasmarek and Lanning-Rush, 2004).

From January 14, 1996, to December 3, 2000, there was about 180 mm of subsidence over an approximately 80-km long arcuate (bow shaped) feature on the northwestern side of Houston, from Sugar Land in southwest Fort Bend County, though Jersey Village in Harris County, to Spring, Tex., in north-central Harris County along the Harris–Montgomery County border (fig. 14). Visible on the interferogram are highly compressed color bands (fringe gradient), which represent a linear feature (lineament). The primary area of subsidence (referred to as the "northwest subsidence feature" or "NWSF") is centered between the cities of Jersey Village and Cypress (fig. 15). Two subsidence features of lesser magnitude exist to the north and south of the NWSF. The Spring subsidence feature (SPSF) extends to the north and east from the NWSF through Spring (fig. 15). The Sugar Land subsidence feature (SLSF) extends from the NWSF through Sugar Land to the south (fig. 15). Additionally, although historical subsidence areas generally are stable or slightly rebounding, measureable subsidence was identified in a 4-km^2 localized area near Seabrook.

The NWSF encompasses an approximately 875 km^2 area. A 0.9-year interferogram from January 30, 1996, to December 30, 1997, showed localized subsidence of greater than 60 mm (> 66 mm/yr) and broad regional (15 to 30 mm/yr) subsidence across the entire NWSF (fig. 16). The subsidence continued in the NWSF between January 14, 1997, and May 19, 1998, with localized subsidence greater than 60 mm (46 mm/yr) (fig. 17),

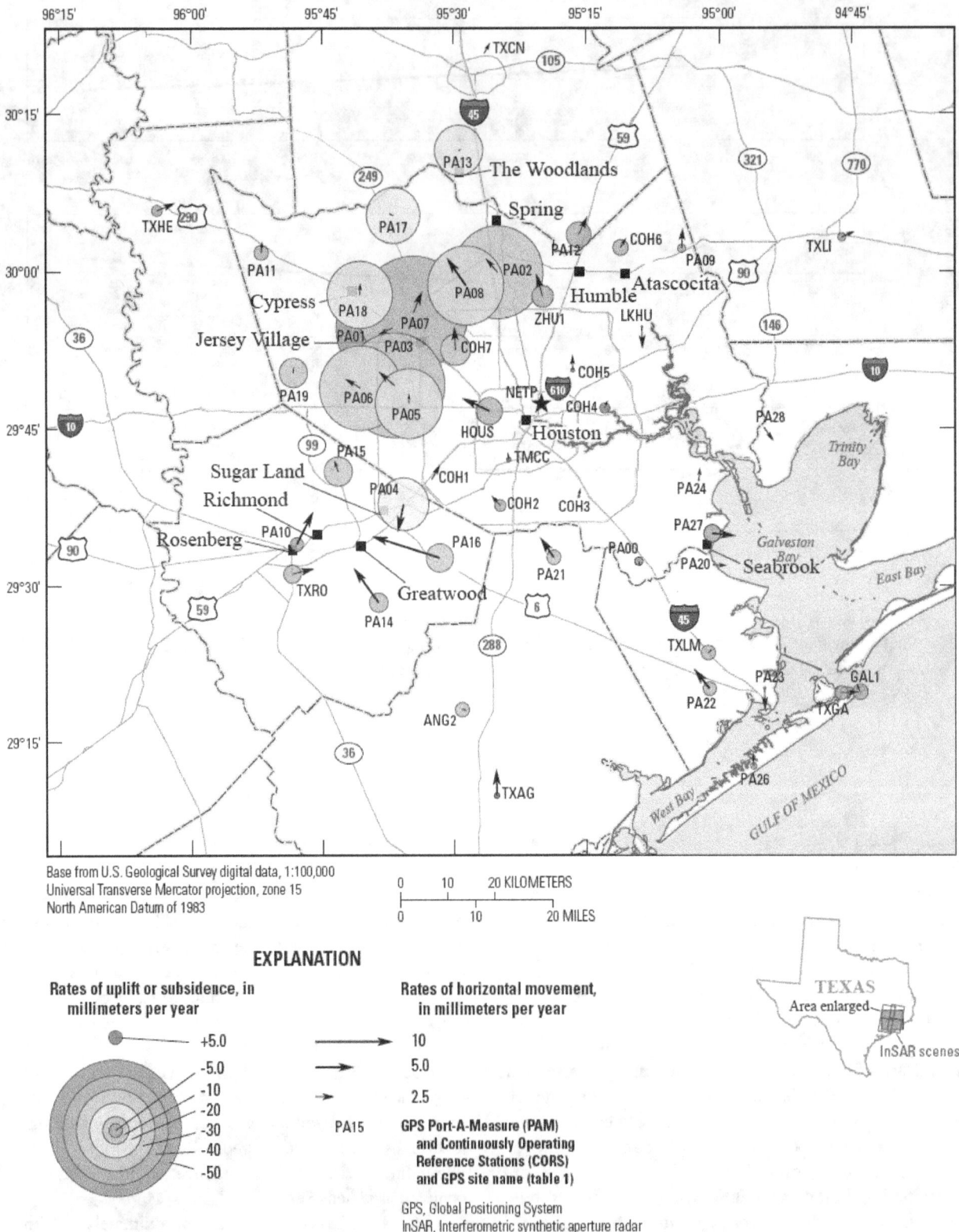

Figure 12. The Global Positioning System- (GPS-) derived horizontal and vertical velocities for Continuously Operating Reference Stations (CORS) and Harris-Galveston Subsidence District Port-A-Measure (PAM) sites, with Northeast 2250 CORS ARP (NETP) (star) as the reference station. The size of the circle is scaled to subsidence magnitude in millimeters per year (mm/yr). The arrows show horizontal motion in millimeters per year of the GPS stations and are oriented towards subsidence or away from uplift.

Table 2. Interferograms generated for the Houston-Galveston region from data obtained during 1993–2000 from European remote sensing (ERS) satellites ERS–1 and ERS–2.

[ID is a unique identifier for an interferogram created from synthetic aperture radar (SAR) master and slave scenes, where the master scene is the beginning time scene that is the reference scene with the year month date format, YYYYMMDD, and the slave scene is the later time SAR scene that is aligned to the master scene with the same year month date format, YYYYMMDD. Track is the orbital track of the satellite. Frame is a latitude and longitude centered area. Scene is the synthetic aperture radar scene area defined by the combination of track and frame]

ID	Master scene	Slave scene	Track	Frame	Scene
1	19960129	19970114	212	2997	T212_F2997
2	19960129	19970325	212	2997	T212_F2997
3	19960129	19971230	212	2997	T212_F2997
4	19960129	19980519	212	2997	T212_F2997
5	19960130	19971230	212	2997	T212_F2997
6	19951226	19960129	212	3015	T212_F3015
7	19951226	19971230	212	3015	T212_F3015
8	19951226	19980901	212	3015	T212_F3015
9	19951226	19981110	212	3015	T212_F3015
10	19960129	19970325	212	3015	T212_F3015
11	19960129	19971230	212	3015	T212_F3015
12	19960129	19980519	212	3015	T212_F3015
13	19960130	19971230	212	3015	T212_F3015
14	19960130	19981110	212	3015	T212_F3015
15	19970114	19980519	212	3015	T212_F3015
16	19970114	19990713	212	3015	T212_F3015
17	19970218	19980203	212	3015	T212_F3015
18	19930902	19960113	484	2997	T484_F2997
19	19960113	19970518	484	2997	T484_F2997
20	19960113	20001203	484	2997	T484_F2997
21	19960114	20001203	484	2997	T484_F2997
22	19930902	19960113	484	3015	T484_F3015
23	19960113	19970518	484	3015	T484_F3015
24	19960113	20000107	484	3015	T484_F3015
25	19960113	20001203	484	3015	T484_F3015
26	19960114	20001203	484	3015	T484_F3015
27	19961229	20001203	484	3015	T484_F3015

and one localized area subsided more than 90 mm between January 29, 1996, and May 19, 1998 (fig. 18). The lineament on the southern boundary of the NWSF may be associated with differential subsidence across the structure or with facies change within the subsurface materials (fig. 18); this lineament is visible in most of the interferograms, including the January 14, 1996, to December 3, 2000, interferogram of the region (fig. 14).

The SPSF encompasses an area approximately 300 km^2 north and east of the NWSF (fig. 15). Ongoing subsidence was observed in the SPSF, with localized subsidence of about 45 mm/yr. The magnitude of subsidence for much of the SPSF is lower that what is observed in the NWSF (fig. 16). Unlike the NWSF, the magnitude and extent of the subsidence in the SPSF did not noticeably change with four additional months included in the interferogram (January 1996 to May 1998) (figs. 16 and 18), which shows that most of the subsidence is apparently west of Humble, Tex., and south of the Harris–Montgomery County line for the time period examined, January 1996 to December 1997. The large InSAR-imaged subsidence signal in the NWSF is also seen in both the horizontal and vertical GPS time-series data (figs. 6 and 19). The GPS-measured subsidence rates are similar in magnitude and match the arcuate pattern seen in the InSAR imagery. Stations that are on the southern side of the area of subsidence are moving northward, into the area with the maximum subsidence. The GPS and InSAR imagery also corresponds in regards to the SPSF at the northeastern end of the NWSF, although subsidence rates are lower and confined to a smaller region.

The SLSF is an open trough encompassing an area approximately 375 km^2 (15 km by 25 km) aligned in a northwest-southeast direction with localized subsidence rates near the center of the trough of approximately 30–40 mm/yr (fig. 15). The northern extent of the SLSF is open-ended and transitions into the much larger NWSF (fig. 15). Between January 1996 and December 2000, there were more than 180 mm of subsidence in the central region of the SLSF, though this amount might not represent the full magnitude of subsidence during this period (fig. 14). In the SLSF area, localized subsidence over the 4-year period resulted in a tight InSAR fringe pattern on the northern side of the trough that exceeded the magnitude of subsidence detectible with InSAR. The full extent of the subsidence on the western margin of the SLSF was difficult to detect with InSAR because the SAR scenes had poor coherency in the vegetated areas west of Richmond, Tex., and Rosenberg, Tex., associated with agriculture production. To minimize the scattering effect from the vegetation cover, temporal decorrelation was applied (a process used to improve coherency, or correlation between the short-term SAR scenes used to create the interferogram). There was also similar agriculture-related decorrelation south of Greatwood, Tex.

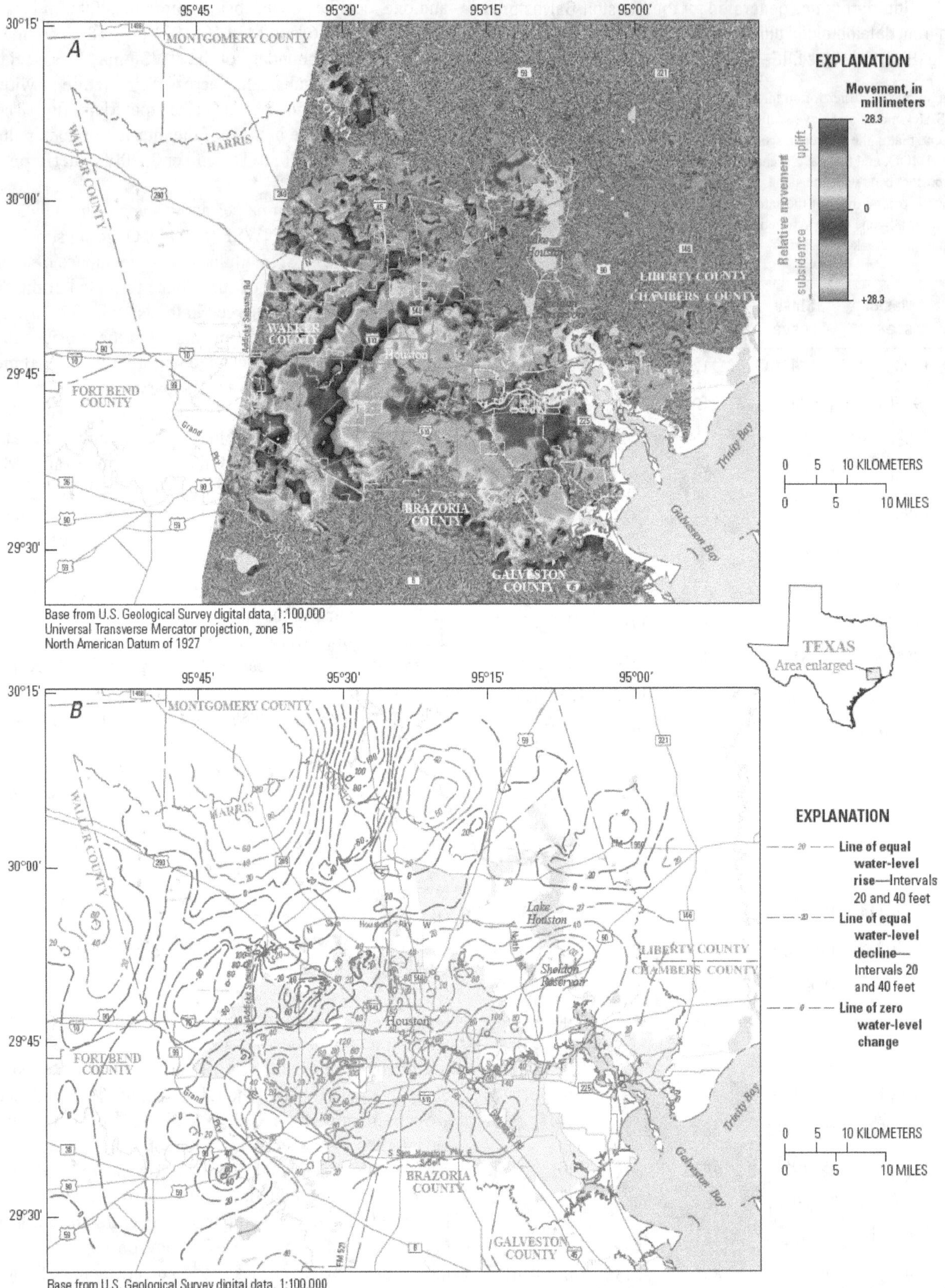

Figure 13.   Comparison of an interferogram with water-level contours for the Houston-Galveston region, Texas. *A*, Interferometric synthetic aperture radar- (InSAR-) measured land subsidence between January 30, 1996, and December 30, 1997. *B*, Evangeline aquifer water-level change contours between 1990 and 2003 (modified from Kasmarek and Lanning-Rush, 2004).

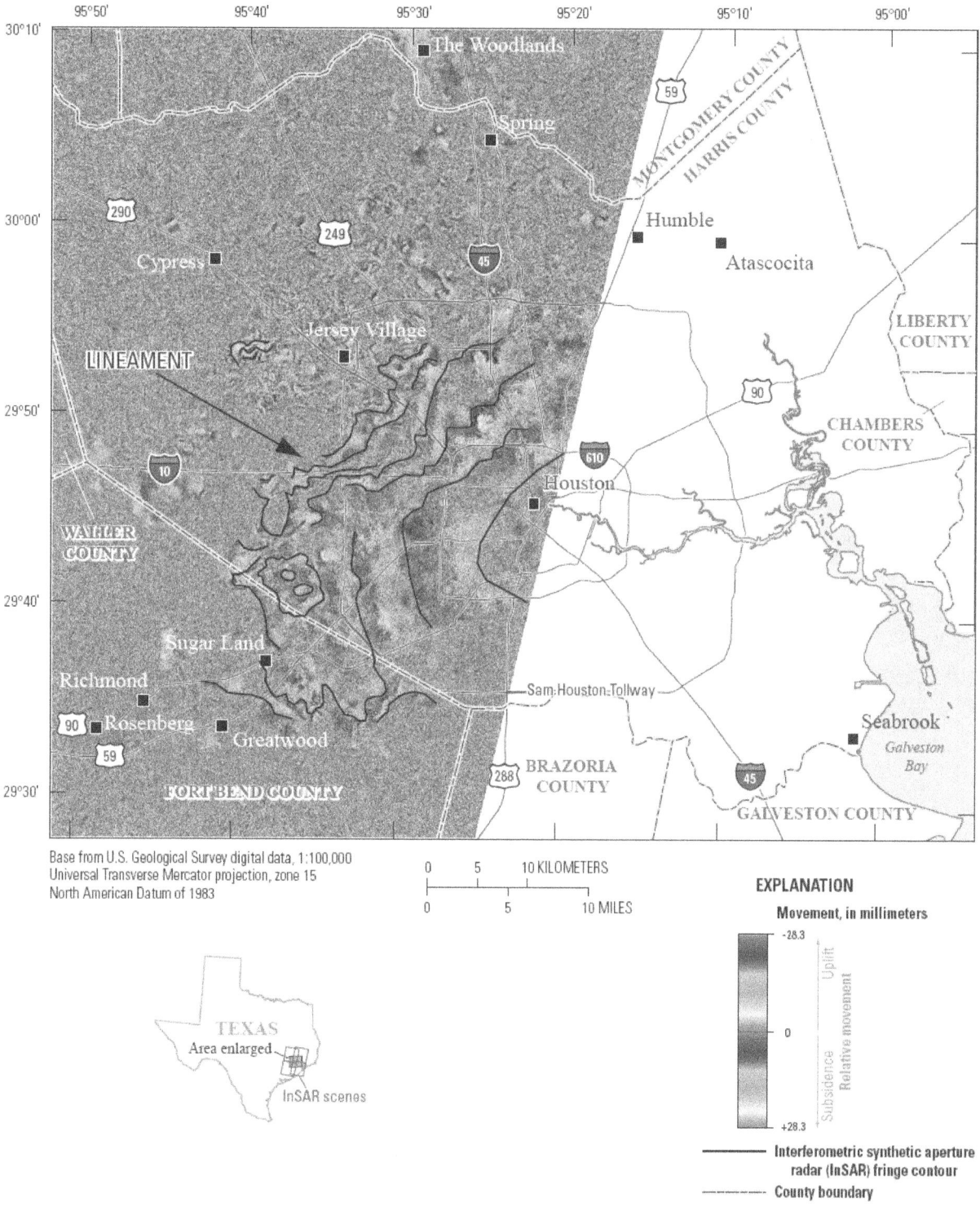

Figure 14.   Interferometric synthetic aperture radar- (InSAR-) measured subsidence from January 14, 1996, to December 3, 2000. Contours delineate the interferogram fringes and show greater than 180 millimeters (mm) of subsidence to the northwest over this 4-year period. The arrow points to highly compressed color bands (fringe gradient), which represent a linear feature (lineament).

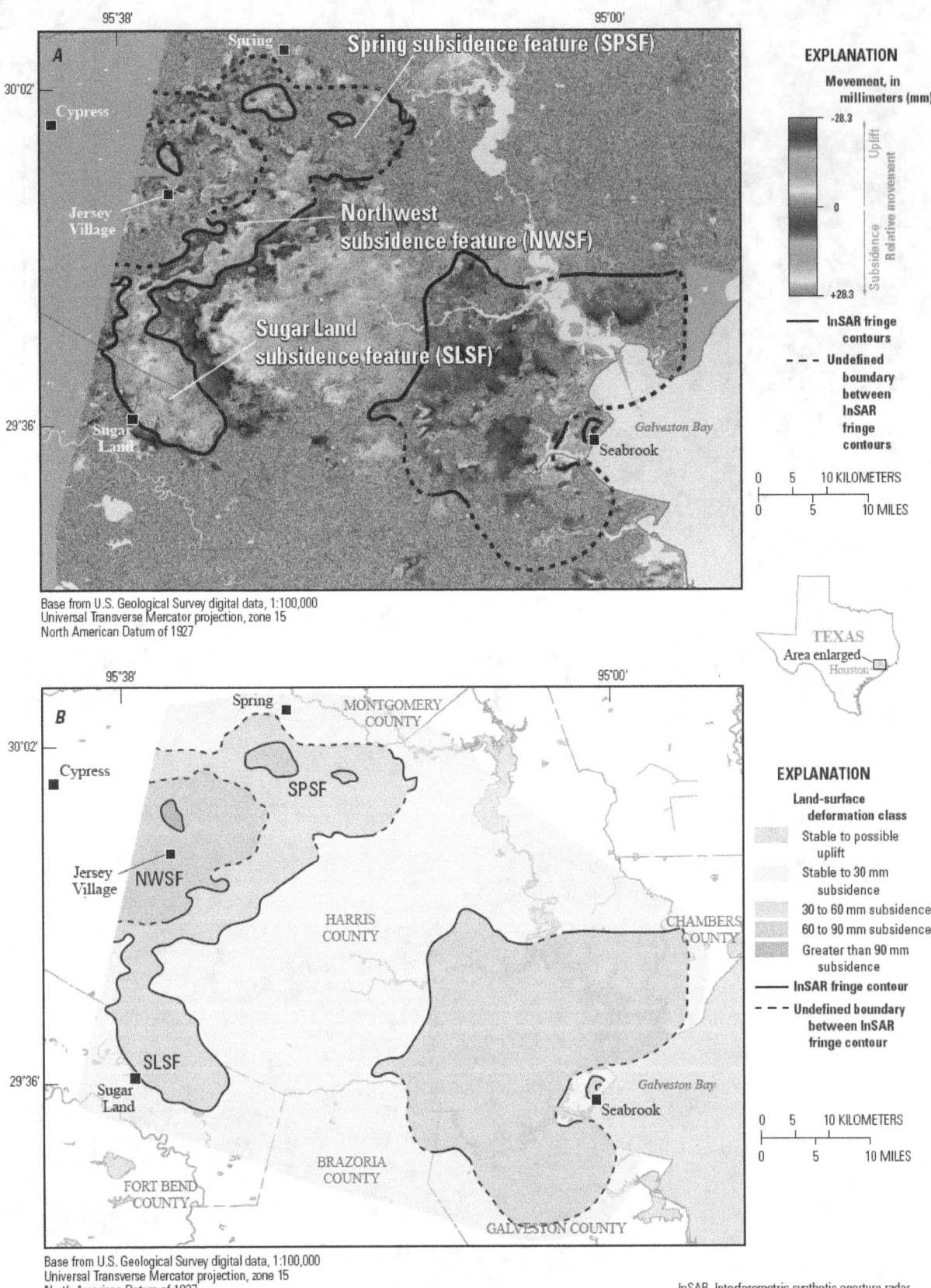

Figure 15. Interferogram for January 30, 1996, to December 30, 1997 showing primary areas of subsidence. *A*, The named subsidence features to the northwest. *B*, Interpolated land-surface deformation contours. The blue- and red-shaded areas represent regions with subsidence or uplift.

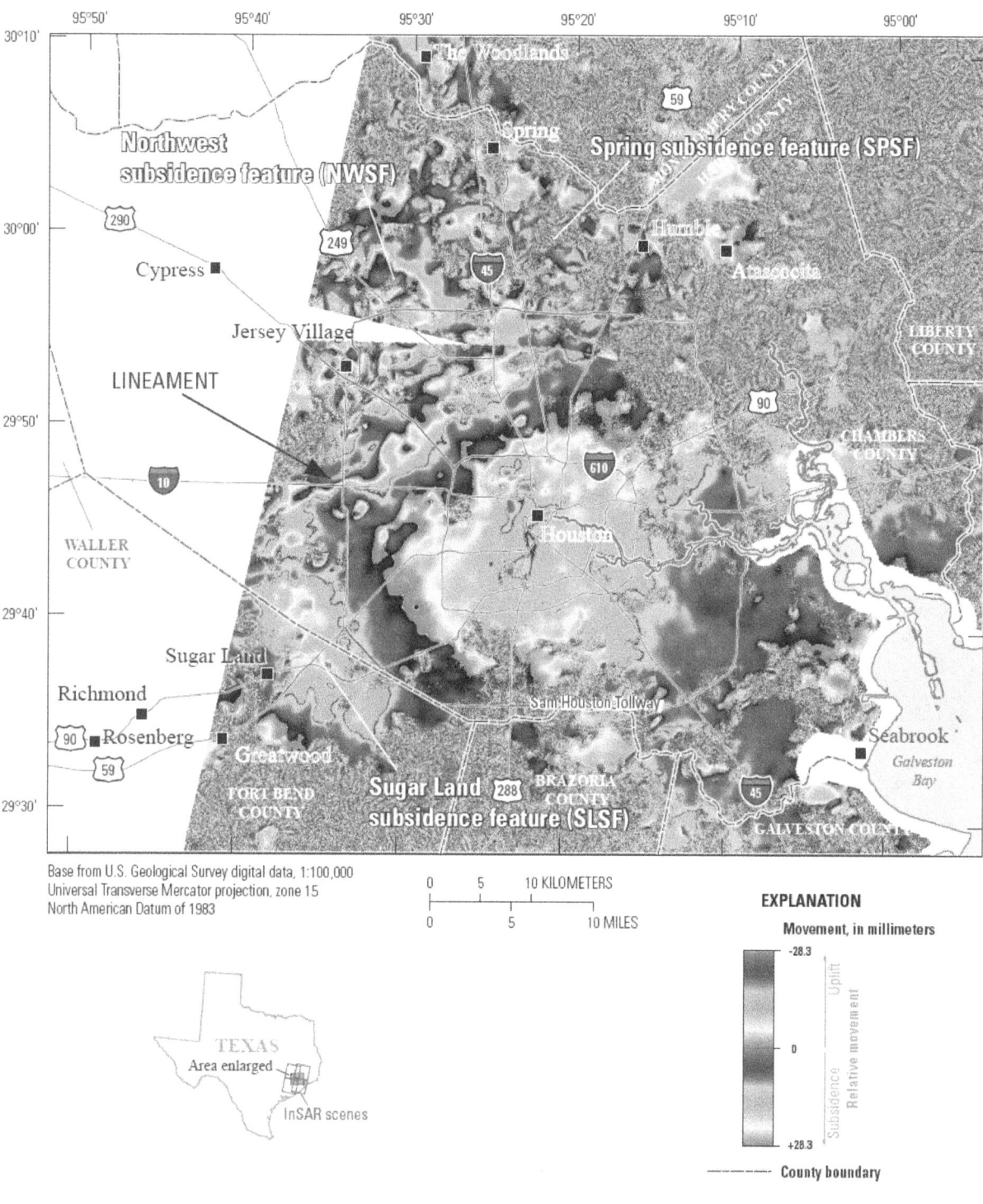

Figure 16. A 0.9-year interferogram (January 30, 1996, to December 30, 1997) of the Houston-Galveston region, Texas, showing 60 millimeters (mm) of subsidence in the northwestern part of the interferogram.

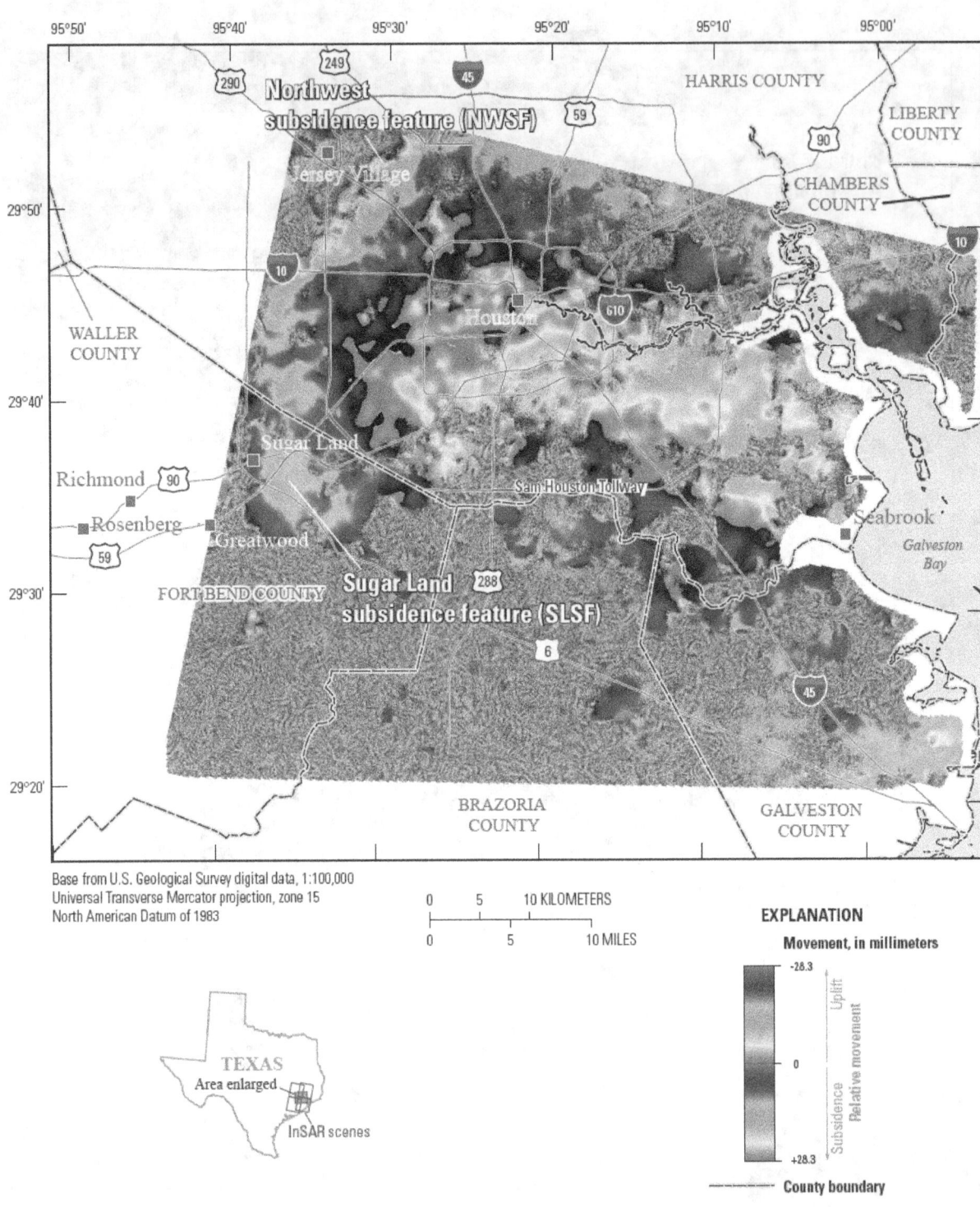

Figure 17. A 1.3-year interferogram (January 14, 1997, to May 19, 1998) of the Houston-Galveston region, Texas, showing approximately 60 millimeters (mm) of subsidence in the northwestern part of the interferogram.

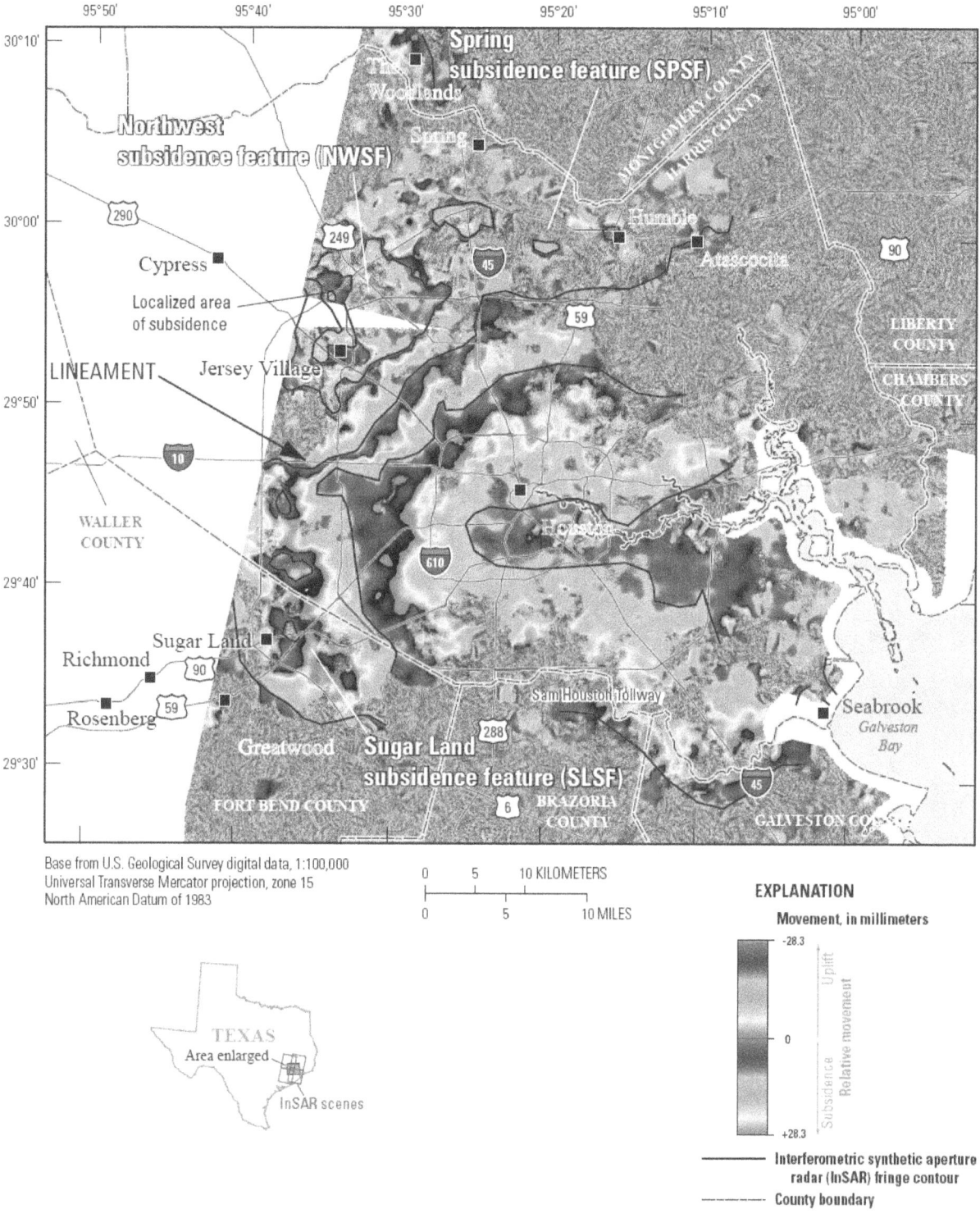

Figure 18. A 2.3-year interferogram (January 29, 1996, to May 19, 1998) of the Houston-Galveston region, Texas, showing a localized area with more than 90 millimeters (mm) of subsidence.

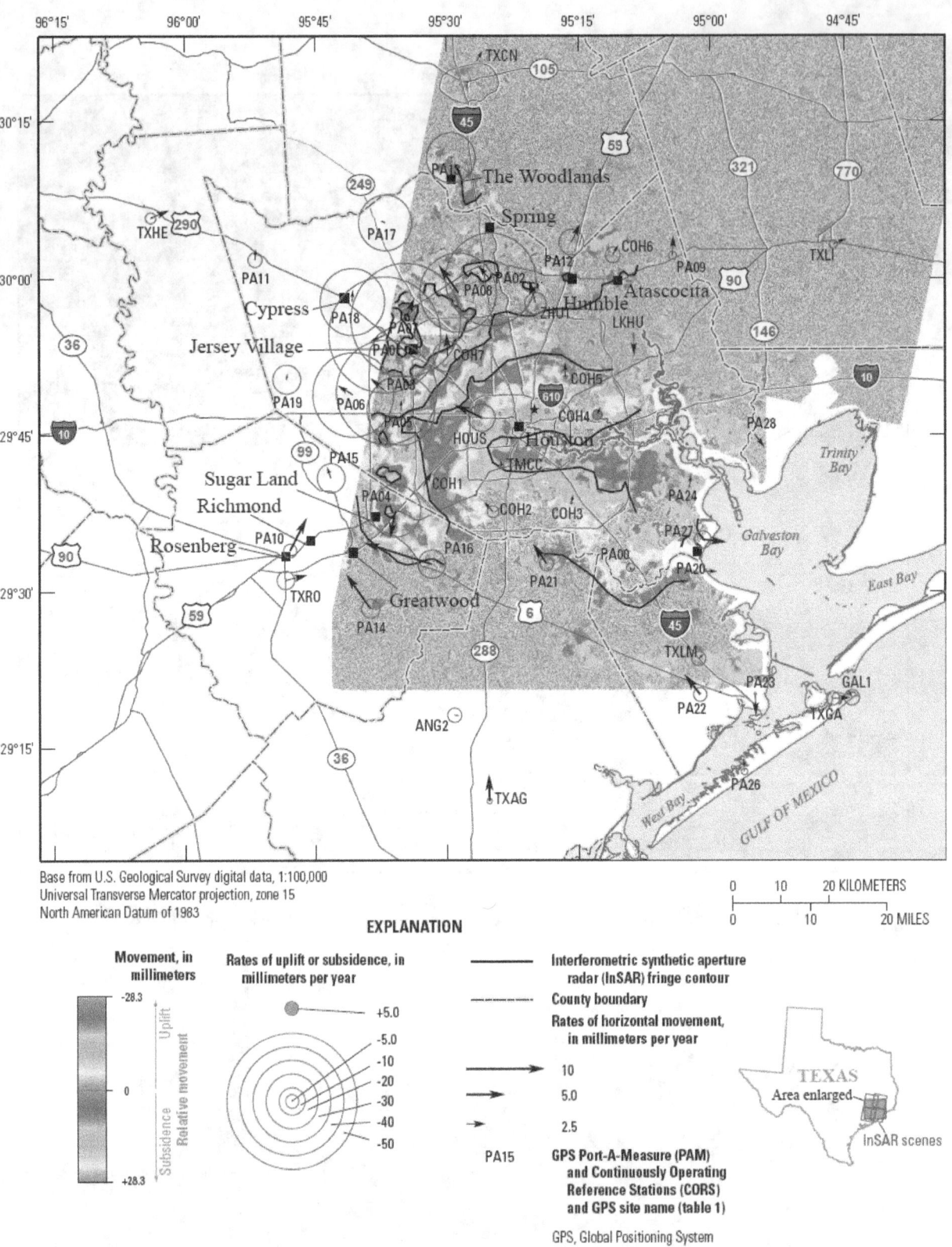

Figure 19.   Comparison of contoured interferograms (January 29, 1996, to May 19, 1998) in the Houston-Galveston region, Texas, with velocities (arrows) in millimeters per year (mm/yr) and subsidence rates (circles) in mm/yr derived from Global Positioning System (GPS) data.

The horizontal and vertical GPS deformation field supports the InSAR-measured subsidence in the SLSF. All of the GPS stations in the area surrounding the SLSF are moving inward towards the subsidence trough, mostly perpendicular to the InSAR fringe gradient. The PAM station, PA04, is north and east of a localized area of subsidence observed from InSAR data (fig. 19), where the interferogram shows about 40 mm of subsidence (19 mm/yr) between January 29, 1996, and May 19, 1998. This amount of subsidence corresponds with the GPS-measured subsidence rate of 19.5 mm/yr, in which the station is moving to the south at a rate of 3.5 mm/yr and has a slightly westward velocity at less than 1 mm/yr directly towards the area with the greatest subsidence. This amount equates to about 290 mm of subsidence between 1994 and 2009 with greater than 50 mm of southward movement of the station. The other stations—PA10, TXRO, PA14, and PA16—also have velocities inward towards the subsidence trough (fig. 19).

Although it is within the recently stable area of historical subsidence, a localized area of subsidence exists near Seabrook. The localized subsidence feature encompasses an area about 8 km^2 where from February 18, 1997, to February 3, 1998, about 28 mm of subsidence occurred (fig. 20). Multiyear interferograms show appreciable (more than 85 mm from January 1996 to December 1997) subsidence in the area with several fringes visible; however, because of decorrelation and the high subsidence magnitudes, it is difficult to determine the maximum subsidence magnitude near Seabrook because many of the fringes are truncated or ill-defined (fig. 21). The subsidence gradient begins northwest of Seabrook with little to no discernible subsidence and extends to the southeast beginning near Taylor Lake Village, Tex., and extending to the area of maximum subsidence near the coast. The PAM GPS station PA27 near Seabrook has an eastward velocity of 3 mm/yr (greater than 20 mm of absolute motion during 2003–9) and is subsiding at 6 mm/yr, or nearly double the eastward motion, matching the InSAR subsidence signal seen in multiple interferograms near Seabrook (fig. 21). The horizontal velocity vectors for PA27 points east towards the area with the maximum localized subsidence (fig. 20).

Limitations

Two primary limitations exist when using the baseline-pair approach for the GPS analyses: (1) all of the motions are relative to the one station and are independent of the regional deformation field, and (2) all local and regional motion measured at the baseline-pair reference station will be mapped into the motion at every other GPS station. The first limitation is not relevant for the scope of this report because it is not necessary to relate the Houston regional subsidence to a presumed stable North American continent or other areas with known subsidence along the Gulf of Mexico Coast. Furthermore, based on the InSAR pixel size (30 m on a side or 900 m^2), there is no advantage to processing the GPS data in a global reference frame to compare with the InSAR imagery. To address the second limitation, an assessment of which CORS station to hold fixed in the baseline-pair analysis is needed to fully characterize the vertical and horizontal motions in the GPS data. The objective is to establish whether or not the reference station is stable. If the station is not stable, then any reference station that is truly subsiding at 10 mm/yr, paired with any other station subsiding at the same rate, would show zero net subsidence. All of the motion is relative to the selected reference station, and if both are subsiding at the same rate, then there is no net difference between the two stations. Continuing with this example, any station that shows uplift of 10 mm/yr, paired with any other station showing uplift at the same rate, would actually have zero net uplift.

InSAR analysis in this study can be limited by a number of factors, some of which are generally described in the "Methodology" section of this report and some of which are described in further detail in appendix 1. Regional agriculture production and dense vegetation, combined with an overall lack of urban reflective surfaces in the western and northern parts of the study area, limited the resolution of the InSAR analysis and any corresponding characterization. Determining the maximum surface motion with InSAR requires an uninterrupted InSAR fringe pattern where the full number of fringes can be uniquely counted, which was not always possible. Incoherency to the northern and eastern parts of the study area decreased the resolution of the InSAR analysis, thereby limiting the subsidence characterization.

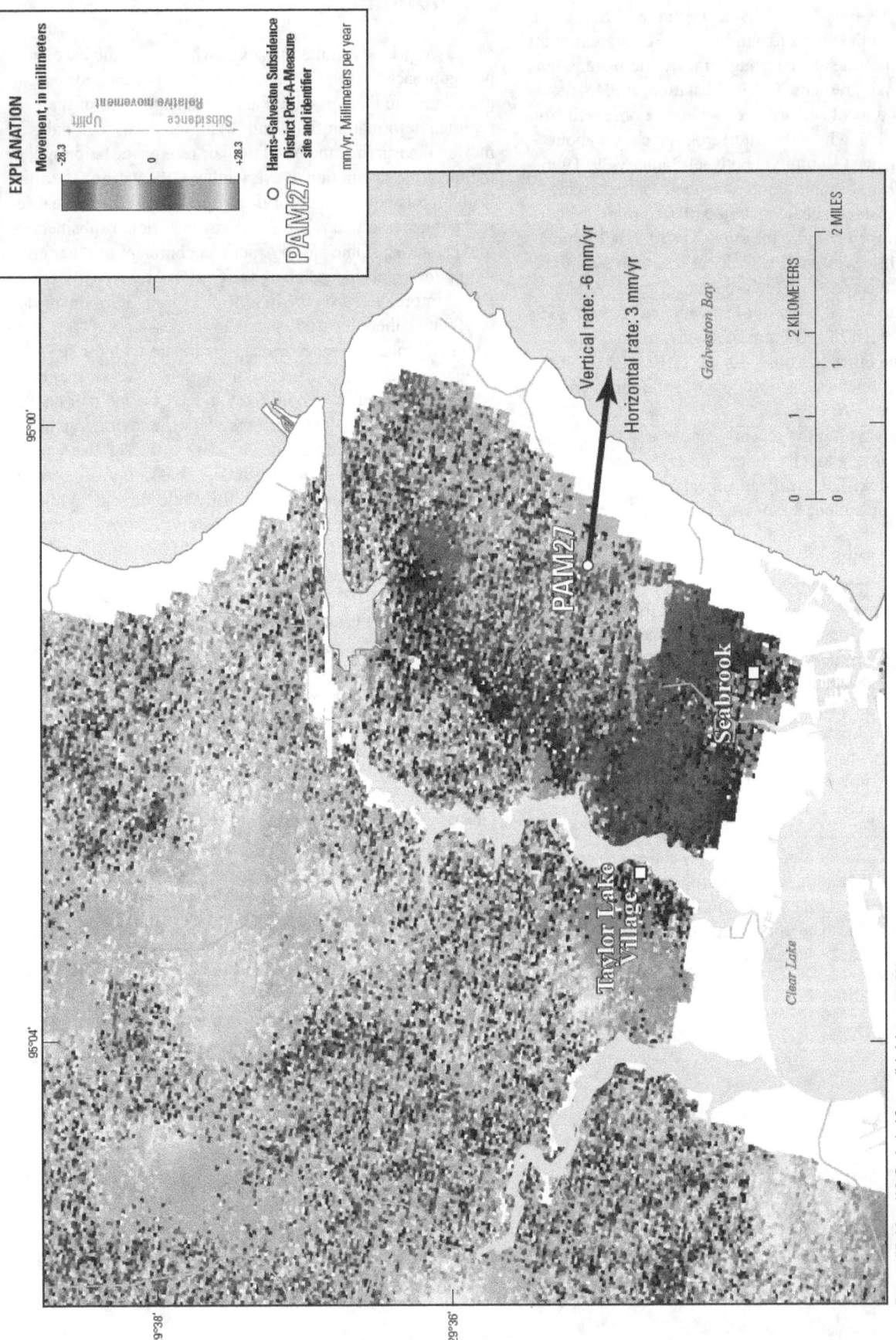

Base modified from U.S. Geological Survey digital data, 1:100,000
Geographic Coordinate System, World Geodetic System 1984

Figure 20.   Detailed interferogram showing subsidence during about 1 year (February 18, 1997, to February 3, 1998) in an area around Seabrook, Texas, with a vector showing the displacement rates in millimeters per year (mm/yr) derived from Global Positioning System (GPS) data.

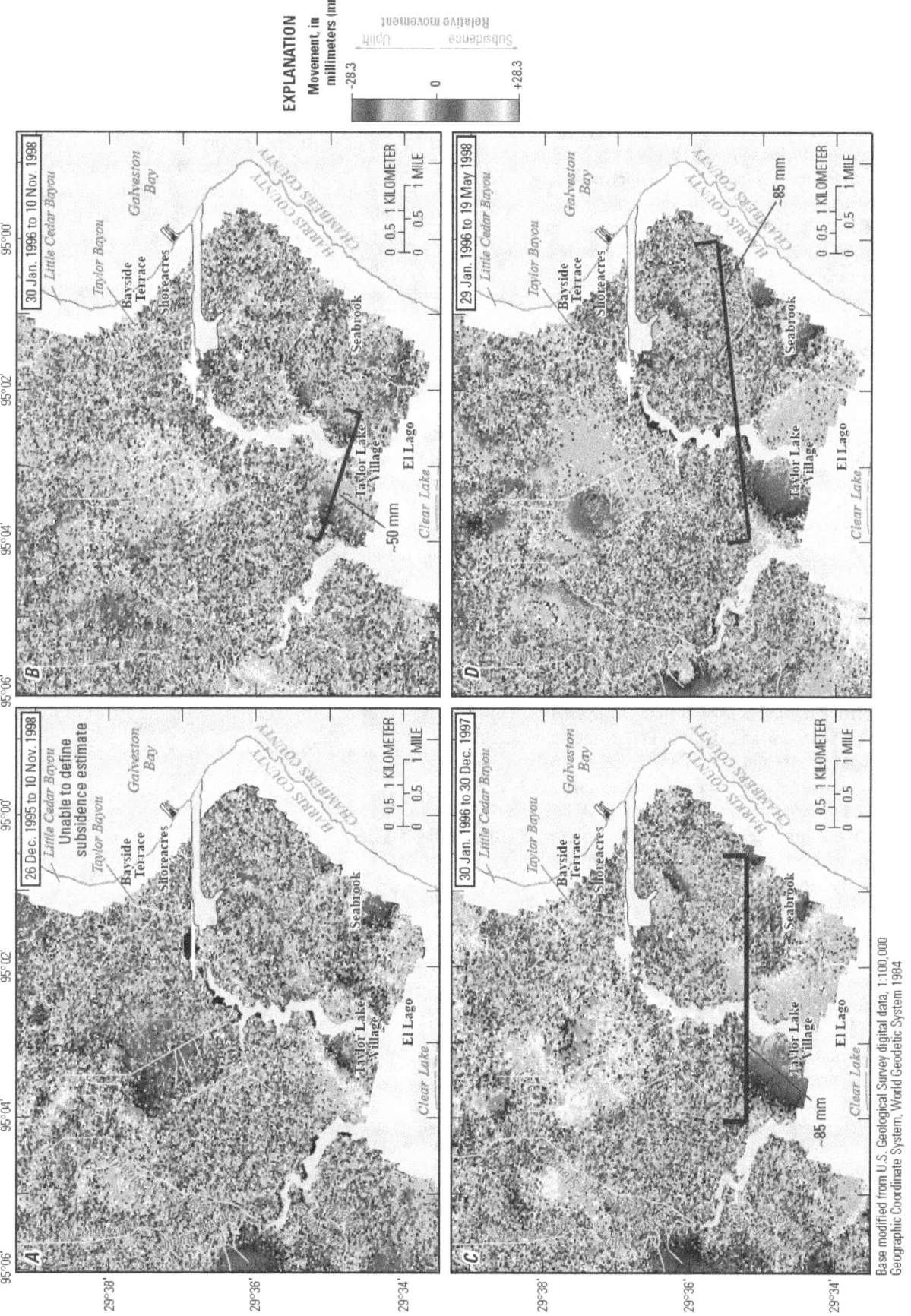

Figure 21. Four detailed interferograms of the Seabrook, Texas, area showing different subsidence results. *A*, Unable to define subsidence estimate. *B*, Approximately 50 millimeters (mm) of subsidence. *C*, Approximately 85 mm of subsidence. *D*, Approximately 85 mm of subsidence.

Summary

Since the early 1900s, groundwater has been the primary source of municipal, industrial, and agricultural water supplies for the Houston-Galveston region, Texas. The region's combination of hydrogeology and nearly century-long use of groundwater has resulted in one of the largest areas of subsidence in the United States; by 1979 as much as 3 meters (m) of subsidence had occurred and approximately 8,300 square kilometers of land had subsided more than 0.3 m. The U.S. Geological Survey, in cooperation with the Harris-Galveston Subsidence District, used interferometric synthetic aperture radar (InSAR) data from two European remote sensing satellites (ERS-1 and ERS-2) to analyze land subsidence in the Houston-Galveston region of Texas. The InSAR data were processed into 27 interferograms that delineate and quantify land subsidence patterns and magnitudes. Contemporaneous data from the Global Positioning System (GPS) data were reprocessed by the National Geodetic Survey and analyzed to support, verify, and provide temporal resolution to the InSAR investigation.

In 2010, the National Geodetic Survey reprocessed all of the regional Continuously Operating Reference Stations (CORS) and Port-A-Measure (PAM) GPS data for the Houston-Galveston region from 1993 through 2009 with respect to the original three CORS stations: Addicks 1795 CORS ARP (antenna reference point) (ADKS), Lake Houston (LKHU), and Northeast 2250 CORS ARP (NETP). A local reference frame was used by holding the CORS station, NETP, fixed; NETP was selected as the reference station in the local baseline-pair reference frame based on the lack of InSAR observed motion at the station and stability in the time-series when compared with LKHU and ADKS. In total, 61 ERS-1 and ERS-2 synthetic aperture radar scenes were acquired, processed, and used to generate 214 interferograms. From the total interferograms produced, only 27 interferograms showed reasonable to good coherency, but many contained atmospheric noise that could not be used to interpret the regional subsidence patterns.

The interferograms show that the area with the maximum historical subsidence has now stabilized and may have a minor uplift component with respect to the surrounding region. Although the historical maximum area of subsidence exists near downtown Houston east along the Houston Ship Channel in southeastern Harris County near Galveston Bay, the InSAR analysis shows that most of the recent subsidence in the area is much farther west of and north of downtown Houston.

The primary area of subsidence determined by InSAR analysis, referred to as the "northwest subsidence feature" (NWSF), is centered between Jersey Village, Tex., and Cypress, Tex. Two subsidence features of lesser magnitude exist to the north and south of the primary area of subsidence near the cities of Spring, Tex. (Spring subsidence feature (SPSF)) and Sugar Land, Tex. (Sugar Land subsidence feature (SLSF)), respectively. Subsidence rates were calculated based on the interval of time between the SAR images. Estimated subsidence in the NWSF is generally 15 to 30 millimeters per year (mm/yr) with localized areas greater than 60 mm/yr, whereas estimated subsidence in localized areas in the SPSF and SLSF is about 45 mm/yr to 30–40 mm/yr, respectively. Multiyear interferograms, near Seabrook, Tex., within the historical subsidence area and nearby Galveston Bay, show several fringes of subsidence (approximately 85 millimeters from January 1996 to December 1997) in the area; however, it is difficult to determine the maximum subsidence magnitude near Seabrook because many of the fringes are truncated or ill-defined. The PAM GPS station, PA27, near Seabrook has an eastward velocity of 3 mm/yr (greater than 20 mm of absolute motion during 2003–9) and is subsiding at 6 mm/yr, or nearly double the eastward motion, matching the InSAR subsidence signal.

Horizontal and vertical GPS velocities are similar in magnitude and match the arcuate pattern seen in the InSAR imagery. Stations that are on the margins of the subsidence features have greater horizontal velocities and are moving toward the areas with the maximum subsidence, mostly perpendicular to the InSAR fringe gradient. Stations on the south side of the SPSF have northern horizontal motion into the subsidence area, and stations bounding the SLSF are pointing radially inward toward the subsidence trough. The PAM station, PA04, which is north and east of the SLSF, has a subsidence a rate of 19.5 mm/yr, in which the station is moving to the south at a rate of 3.5 mm/yr, and has a slightly westward velocity at less than 1 mm/yr, which is directly towards the area with the greatest subsidence. This equates to about 290 mm of subsidence between 1994 and 2009 with greater than 50 mm of southward movement of the station; similar patterns are observed with the other GPS stations.

References Cited

Amelung, Falk, Galloway, D.L., Bell, J.W., Zebker, H.A., and Laczniak, R.J., 1999, Sensing the ups and downs of Las Vegas—InSAR reveals structural control of land subsidence and aquifer-system deformation: Geology, v. 27, p. 483–486.

Baker, E.T., 1979, Stratigraphic and hydrogeologic framework of part of the Coastal Plain of Texas: Texas Department of Water Resources Report 236, 43 p.

Bawden, G.W., Sneed, Michelle, Stork, S. V., and Galloway, D. L., 2003, Measuring human-induced land subsidence from space: U.S. Geological Survey Fact Sheet 069–03, 4 p.

Bawden, G.W., Thatcher, Wayne, Stein, R.S., Hudnut, K.W., and Peltzer, Gilles, 2001, Tectonic contraction across Los Angeles after removal of groundwater pumping effects: Nature, v. 412, p. 812–815.

Blewitt, Geoffrey, Bock, Yehuda, and Kouba, Jan, 1995, Constructing the IGS polyhedron by distributed processing *in* Zumberge, J.F. and Liu, R., eds., IGS Workshop Proceedings—Densification of the IERS terrestrial reference frame through regional GPS networks, November 30–December 2, 1994, Pasadena, California: International GNSS Service, p. 21–38.

Borik, Milan, and Novotny, Jaroslav, 2011, Terrain deformations in the area of the Ervenice Corridor by 2-pass method, *in* Halounová, Lena , ed., Remote Sensing and Geoinformation: European Association of Remote Sensing Laboratories, p. 517–521, accessed June 12, 2012, at http://www.earsel.org/symposia//2011-symposium-Prague/Proceedings/PDF/Radar%20Remote%20Sensing/59%20ok25-a2430-borik.pdf.

Buckley, S.M., Rosen, P.A., Hensley, Scott, and Tapley, B.D., 2003, Land subsidence in Houston, Texas, measured by radar interferometry and constrained by extensometers: Geophysical Research, v. 108, no. B11, 18 p., accessed February 9, 2012, at http://www.agu.org/journals/jb/jb0311/2002JB001848/2002JB001848.pdf.

Byun, S.H., Hajj, G.A., and Young, L.E., 2002, Assessment of GPS signal multipath interference, *in* Proceedings of the 2002 National Technical Meeting of The Institute of Navigation, San Diego, Calif., January 2002: The Institute of Navigation, p. 694–705.

Catchings, R.D., Goldman, M.R., Gandhok, Gini., Rymer, M.J., and Underwood, D.H., 2000, Seismic imaging evidence for faulting across the northwestern projection of the Silver Creek Fault, San Jose, California: U.S. Geological Survey Open File Report 00–0125, 29 p.

Chowdhury, A.H., and Turco, M.J., 2006, Geology of the Gulf Coast aquifer, Texas, *in* Mace, R.E., Davidson, S.C., Angle, E.S., and Mullican, W.F., eds., Aquifers of the gulf coast of Texas: Texas Water Development Board Report 365, chap. 2, p. 23–50, accessed March 7, 2011, at http://www.twdb.state.tx.us/publications/reports/numbered_reports/doc/R365/ch02-Geology.pdf.

Coplin, L.S., and Galloway, Devin, 1999, Houston-Galveston, Texas—Managing coastal subsidence, *in* Galloway, Devin, Jones, D.R., and Ingebritsen, S.E., eds., Land subsidence in the United States: U.S. Geological Survey Circular 1182, p. 35–48.

European Space Agency, 2007, InSAR Principles—Guidelines for SAR interferometry processing and interpretation (ESA TM-19): The Netherlands, European Space Agency, 48 p.

European Space Agency, 2012, Missions—Observing the earth: accessed June 12, 2012, at http://www.esa.int/esaEO/SEMGWH2VQUD_index_0_m.html.

Fort Bend Subsidence District, 2009, Fort Bend Subsidence District 2003 regulatory plan [amended 2007, 2009]: 14 p., accessed March 7, 2011, at http://www.fbsubsidence.org/assets/pdf/FBRegPlan.pdf.

Gabrysch, R.K., and Neighbors, Ronald, 2005, Measuring a century of subsidence in the Houston-Galveston region, Texas, USA, *in* Proceedings of the Seventh International Symposium on Land Subsidence, Shanghai, PR China, October 23–28: p. 379–387.

Galloway, D.L., and Hoffmann, Jörn, 2007, The application of satellite differential SAR interferometry-derived ground displacements in hydrogeology: Hydrogeology Journal, v. 15, no. 1, p. 133–154.

Galloway, D.L., Hudnut, K.W., Ingebritsen, S.E., Phillips, S.P., Peltzer, Gilles, Rogez, F., and Rosen, P.A., 1998, Detection of aquifer system compaction and land subsidence using interferometric synthetic aperture radar, Antelope Valley, Mojave Desert, California: Water Resources Research, v. 34, p. 2573–2585.

Galloway, Devin, Jones, D.R., and Ingebritsen, S.E., eds., 1999, Land subsidence in the United States: U.S. Geological Survey Circular 1182, 177 p.

Galloway, Devin, Jones, D.R., and Ingebritsen, S.E., 2000, Measuring land subsidence from space: U.S. Geological Survey Fact Sheet 051–00, 4 p.

GAMMA Remote Sensing and consulting, 2012, SAR and interferometry software: accessed February 1, 2012, at http://www.gamma-rs.ch/.

Harris-Galveston Subsidence District, 2010, District regulatory plan 1999 [amended 2001, 2010]: 16 p., accessed March 7, 2011, at http://www.hgsubsidence.org/assets/pdfdocuments/HGRegPlan.pdf.

Harris-Galveston Subsidence District, 2011, Subsidence charts: accessed June 18, 2011, at http://mapper.subsidence.org/Chartindex.htm.

Harris-Galveston Subsidence District, 2012, GPS—Using technology from the world above to monitor the land below: accessed June 7, 2012, at http://www.hgsubsidence.org/about/subsidence/gps.html.

Helsel, D.R., and Hirsch, R.M., 2002, Statistical methods in water resources—Hydrologic analysis and interpretation: Techniques of Water-Resources Investigations of the U.S. Geological Survey, book 4, chap. A3, 510 p.

Ikehara, M.E., Galloway, D.L., Fielding, E.J., Bürgmann, Roland, Lewis, A.S., and Ahmadi, B., 1998, InSAR imagery reveals seasonal and longer-term land-surface elevation changes influenced by ground-water levels and fault alignment in Santa Clara Valley, California [abs.]: EOS Transactions, American Geophysical Union, no. 45, p. F37.

Johnson, M.R., Ramage, J.K., and Kasmarek, M.C., 2011, Water-level altitudes 2011 and water-level changes in the Chicot, Evangeline, and Jasper aquifers and compaction 1973–2010 in the Chicot and Evangeline aquifers, Houston–Galveston region, Texas: U.S. Geological Survey Scientific Investigations Map 3174, 17 p., 16 sheets.

Kasmarek, M.C., Gabrysch, R.K., and Johnson, M.R., 2009a, Estimated land-surface subsidence in Harris County, Texas, 1915–17 to 2001: U.S. Geological Survey Scientific Investigations Map 3097, 2 sheets. (Also available at http://pubs.usgs.gov/sim/3097/.)

Kasmarek, M.C., and Houston, N.A., 2007, Water-level altitudes 2007 and water-level changes in the Chicot, Evangeline, and Jasper aquifers and compaction 1973–2006 in the Chicot and Evangeline aquifers, Houston-Galveston region, Texas: U.S. Geological Survey Scientific Investigations Map 2968, 159 p., 18 sheets.

Kasmarek, M.C., Houston, N.A., and Ramage, J.K., 2009b, Water-level altitudes 2009 and water-level changes in the Chicot, Evangeline, and Jasper aquifers and compaction 1973–2008 in the Chicot and Evangeline aquifers, Houston-Galveston Region, Texas: U.S. Geological Survey Scientific Investigations Map 3081, 3 p., 16 sheets. (Also available at http://pubs.usgs.gov/sim/3081/.)

Kasmarek, M.C., Johnson, M.R., and Ramage, J.K., 2010, Water-level altitudes 2010 and water-level changes in the Chicot, Evangeline, and Jasper aquifers and compaction 1973–2009 in the Chicot and Evangeline aquifers, Houston-Galveston region, Texas: U.S. Geological Survey Scientific Investigations Map 3138, 17 p., 16 sheets, 1 appendix.

Kasmarek, M.C., and Lanning-Rush, Jennifer, 2004, Water-level altitudes 2003 and water-level changes in the Chicot, Evangeline, and Jasper aquifers and compaction 1973–2002 in the Chicot and Evangeline Aquifers, Houston-Galveston Region, Texas: U.S. Geological Survey Open-File Report 2003–109, 15 figs.

Kasmarek, M.C., and Robinson, J.L., 2004, Hydrogeology and simulation of ground-water flow and land-surface subsidence in the northern part of the Gulf Coast aquifer system, Texas: U.S. Geological Survey Scientific Investigations Report 2004–5102, 111 p.

Kreps, M.A., 1987, Study links subsidence and flooding in inland areas: Houston Chronicle archives, accessed June 7, 2012, at http://www.chron.com/CDA/archives/archive.mpl/1987_472739/study-links-subsidence-and-flooding-in-inland-area.html.

Landtech Consultants, Inc., 2003, Tropical storm Allison recovery DR-1379 Harris County Texas—Benchmark control network technical report Addicks Dam, Cypress Creek, Spring Creek, White Oak Bayou, and Willow Creek: accessed June 15, 2012, at http://hcedweb3.eng.hctx.net/techdocs/Report_3.pdf.

Mateus, Pedro, Nico, Giovanni., and Catalão, João, 2011, Can spaceborne SAR interferometry be used to study the temporal evolution of PWV?: Atmospheric Research ISSN 0169-8095, 10.1016/j.atmosres.2011.10.002.

National Geodetic Survey, 2011, Continuously Operating Reference Station (CORS): accessed June 28, 2011, at http://www ngs.noaa.gov/CORS/.

National Oceanic and Atmospheric Administration, 2008, Comparative climatic data for the United States through 2008: National Climatic Data Center, accessed June 8, 2010 at http://ols nndc noaa.gov/plolstore/plsql/olstore.prodspecific?prodnum=C00095-PUB-A0001#OVERVIEW.

Paine, J.G., 1993, Subsidence of the Texas coast–Inferences from historical and late Pleistocene seal levels: Tectonophysics, v. 222, p. 445–458.

Pratt, W.E., and Johnson, D.W., 1926, Local subsidence of the Goose Creek oil field: Journal of Geology, v. 34, p. 577–590.

Rothacher, Markus, 2001, Comparison of absolute and relative antenna phase center variations: GPS Solutions, v. 4, no. 4, p. 55–60.

Ryder, P.D., and Ardis, A.F., 2002, Hydrology of the Texas Gulf Coast aquifer systems: U.S. Geological Survey Professional Paper 1416-E, 77 p., 8 plates. (Also available at http://pubs.usgs.gov/pp/1416e/report.pdf.)

Schmidt, D.A., and Bürgman, Roland, 2003, Time-dependent land uplift and subsidence in the Santa Clara valley, California, from a large interferometric, synthetic aperture radar dataset: Geophysical Research, v. 108, no. B9, 13 p., accessed February 9, 2012, at http://www.agu.org/journals/jb/jb0309/2002JB002267/2002JB002267.pdf.

Seifert, John, and Drabek, Christopher, 2006, History of production and potential future production of the Gulf Coast Aquifer, chap. 16 of Mace, R.E., Davidson, S.C., Angle, E.S., and Mullican, W.F. III, eds., Aquifers of the gulf coast of Texas: Texas Water Development Board Report 365, p. 261–271.

Snay, R.A., Soler, Tomás, and Eckl, Mark, 2002, GPS precision with carrier phase observations—Does distance and/or time matter?: Professional Surveyor, v. 22, no. 10, p. 20, 22, 24.

Sneed, Michelle, and Brandt, Justin, 2007, Detection and measurement of land subsidence using Global Positioning System surveying and interferometric synthetic aperture radar, Coachella Valley, California, 1996–2005: U.S. Geological Survey Scientific Investigations Report 2007–5251, 31 p.

Soler, Tomás, Johnson, R.E., Thormahlen, L.F., and Foote, R.H., 2001, Combining two GPS techniques—Parting the waters: GPS World, v. 12, no. 5, p. 28–31.

Soler, Tomás, and Marshall, John, 2002, Rigorous transformation of variance-covariance matrices of GPS-derived coordinates and velocities: GPS Solutions, v. 6, no. 1–2, p. 76–90.

Stork, S.V., and Sneed, Michelle, 2002, Houston-Galveston Bay Area, Texas, from space—A new tool for mapping land subsidence: U.S. Geological Survey Fact Sheet 110–02, 4 p.

U. S. Census Bureau, 2001, Census 2000 brief—Population change and distribution, 1990–2000: accessed June 7, 2012, at http://www.census.gov/prod/2001pubs/c2kbr01-2.pdf.

U.S. Census Bureau, 2009a, Annual estimates of the resident population for the United States, regions, States, and Puerto Rico—April 1, 2000 to July 1, 2008 (NST-EST2008-01): U.S. Census Bureau, Population Division, accessed June 12, 2012, at http://www.census.gov/popest/data/historical/2000s/vintage_2008/index.html.

U.S. Census Bureau, 2009b, Cumulative estimates for metropolitan statistical areas and rankings—April 1, 2000 to July 1, 2008 (CBSA-EST2008-07): U.S. Census Bureau, Population Division, accessed June 12, 2012, at http://www.census.gov/popest/data/historical/2000s/vintage_2008/metro.html.

Verbeek, E.R., Ratzlaff, K.W., and Clanton, U.S., 1979, Faults in parts of north-central and western Houston metropolitan area, Texas: U.S. Geological Survey Miscellaneous Field Studies Map 1136, 2 plates. (Also available at http://pubs.water.usgs.gov/mf1136/.)

Wei, Meng, and Sandwell, D.T., 2010, Decorrelation of L-Band and C-Band interferometry over vegetated areas in California: IEEE Transactions on Geoscience and Remote Sensing, v. 28, no. 7, p. 2,942–2,952, accessed February 9, 2012, at http://adsabs.harvard.edu/abs/2010ITGRS..48.2942W.

Appendix 1—Interferometic Synthetic Aperture Radar Processing and Error Analysis

Introduction

The processing of satellite synthetic aperture radar (SAR) scenes to generate interferograms is a multistep process that transforms the initial radar echoes emitted by the satellite into georeferenced radar images (interferograms) that can detect subcentimeter changes in the land surface position. This appendix outlines the steps used to process the interferograms: SAR scene focusing, creating an interferogram, georeferencing, removing topographic effects, and unwrapping the interferogram to determine the absolute magnitude of the land surface motion. Limitations and potential error sources associated with the analysis and interpretation of interferograms are also discussed (Schmidt and Bürgman, 2003; European Space Agency, 2007).

SAR Scenes Focus and Alignment

The first step to creating an interferometic synthetic aperture radar (InSAR) image is to focus and align two SAR scenes collected along the same orbit trajectory (same point in space) on different orbits; for optimal results, the orbital path between each pass should be closer than 200 meters (m). Raw SAR data typically consists of 28,000 lines of data, where each line is a radar echo that was emitted from the satellite; traveled to the ground surface; interacted with the land, vegetation, buildings, and so forth; and returned to the satellite. Image focusing combines individual radar echoes into one coherent image approximately 100- by 100-kilometers (km) in size for European remote sensing (ERS) satellites ERS-1 and ERS-2 that is typically shown as a grey-scale image with increased intensity correlating to greater amount of signal reflected back to the satellite (fig. 1.1). Once the scenes have been focused, the two scenes need to be precisely aligned. This is accomplished by employing a pixel-correlated routine that matches common pixel patterns between the master scene (the scene that is held fixed, typically the scene with the earliest date) (fig. 1.1*A*) and the slave scene (any scene that is aligned to the master, typically the scene with the latest data) (fig. 1.1*B*). The orbital ephemerides (orbit position information for the satellite during the scene acquisition) for each scene provide the initial alignment position, and then the pixel correlation routine aligns each image so that the location of each pixel in the slave scene matches the master (European Space Agency, 2007).

Creating an Interferogram

The next stage of the process is to generate an interferogram from the aligned SAR scenes obtained from ERS-1 and ERS-2. A pulse of energy is emitted from each satellite that propagates through the atmosphere, interacts with the ground, and bounces back to the satellite where three pieces of information are recorded: (1) the two-way travel time from the satellite to the ground and back; (2) the amount of signal bounced back, known as magnitude or intensity; and (3) the phase of the return signal, which is described below. These data are used to generate the interferogram. The travel time is the time it takes the pulse to travel from the satellite to a point on the ground and then back to the satellite and is used to calculate the distance between the satellite and the ground (Schmidt and Bürgman, 2003; European Space Agency, 2007; Sneed and Brandt, 2007). Since the signal travels the distance twice, once from the satellite to the ground and once from the ground back to the satellite, the true distance is half, or divided by two. Distance is calculated by:

$$D = (t * c) / 2 \qquad (1.1)$$

where

D	is the distance from the satellite to the ground surface, in meters
t	is the travel time of the pulse from the satellite to the ground and back to the satellite, in seconds, and
c	is the speed of light, in meters per second.

The magnitude of return signal is a measure of the amplitude, intensity, or radar albedo returned for that point and is often visualized as a grey-scale photo-like image of the land surface. The phase is the fraction of a wavelength that is returned to the satellite. ERS-1 and ERS-2 are C-band (56 millimeters (mm)) radar satellites. The distance between the ground and the satellite can be quantified with a very large number of full sinusoidal wavelengths (satellite altitude * 56 mm), plus the fractional wavelength distance, to fully reach the ground from the satellite. If the ground surface was stable between the two satellite acquisitions, then the phase would be the same for both scenes. If subsidence occurred between the two satellite acquisition dates, however, then there would be a phase shift, in this case a phase lag (a positive change in phase between the master and slave scenes) corresponding to an increase in the distance between the ground and satellite. Alternatively, if there were uplift, then the phase shift would be negative (a phase advance) and correspond to a decrease in the distance between the ground and satellite (European Space Agency, 2007).

The interferogram is produced by comparing the change in phase for each corresponding pixel in the aligned SAR scenes. At this stage in the process, the interferogram is a non-georeferenced phase-change image that identifies changes in land surface elevation, positive or negative, by repeating a pattern of colors (fig. 1.2). Each unique pattern, or fringe, represents one sinusoidal wavelength (0 to 2π: 0–28.3 millimeters (mm) for ERS-1 and ERS-2 C-band radar satellites) (fig. 1.2) (European Space Agency, 2007).

Georeferencing the Interferogram

The interferogram is georeferenced by aligning the master SAR scene to a digital elevation model (DEM) using the same pixel correlation routine described above. A synthetic

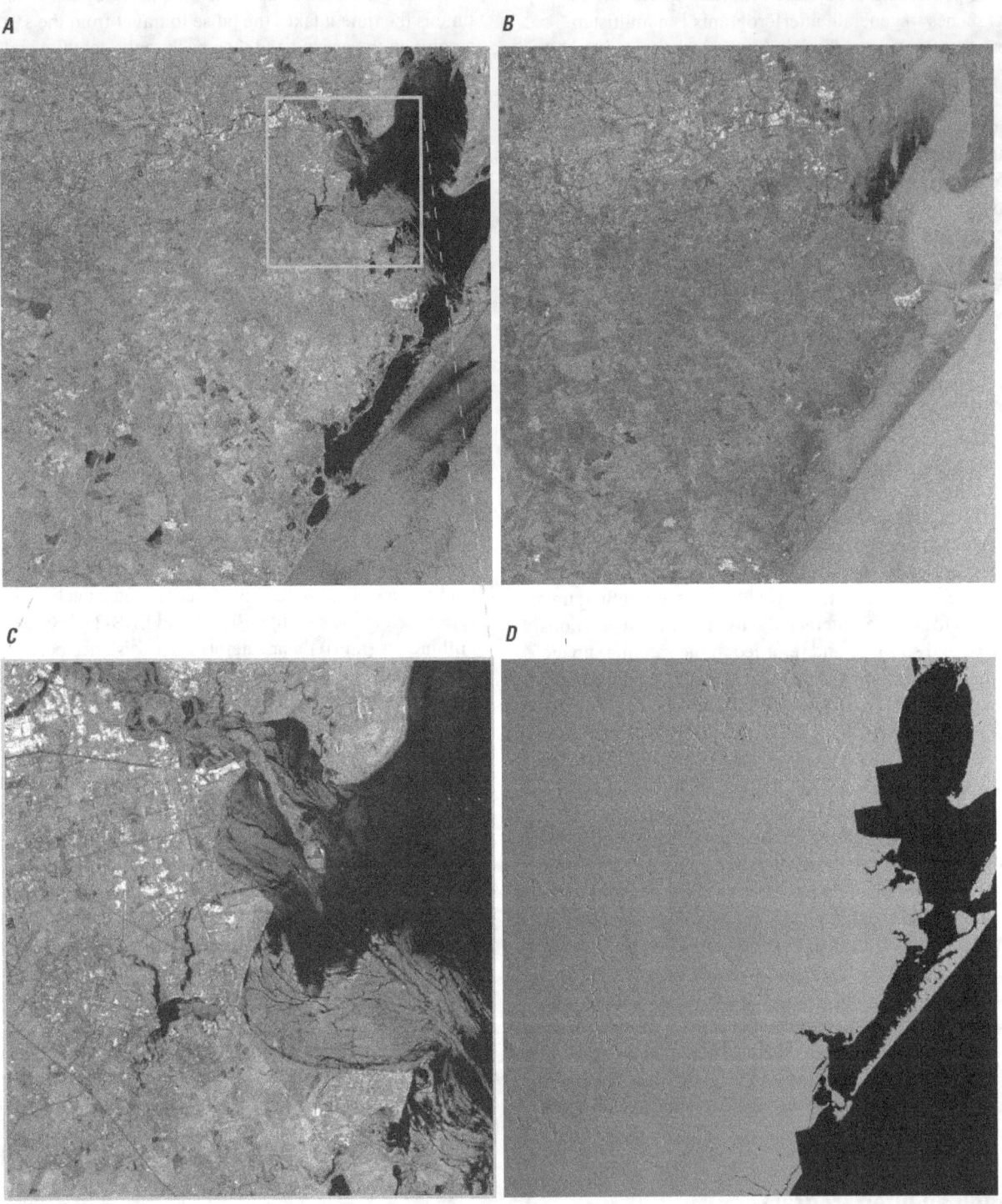

Figure 1.1.   Example synthetic aperture radar (SAR) scenes for the Houston-Galveston region where increased pixel intensity correlates to greater radar signal returned to the satellite: *A*, SAR scene collected on December 30, 1997; *B*, SAR scene collected on July 28, 1998; *C*, Zoom-in region in the box on *A*; *D*, SAR scene from a digital elevation model (DEM) using the orbital position from scene *A*.

A. Simplified interferogram showing uplift

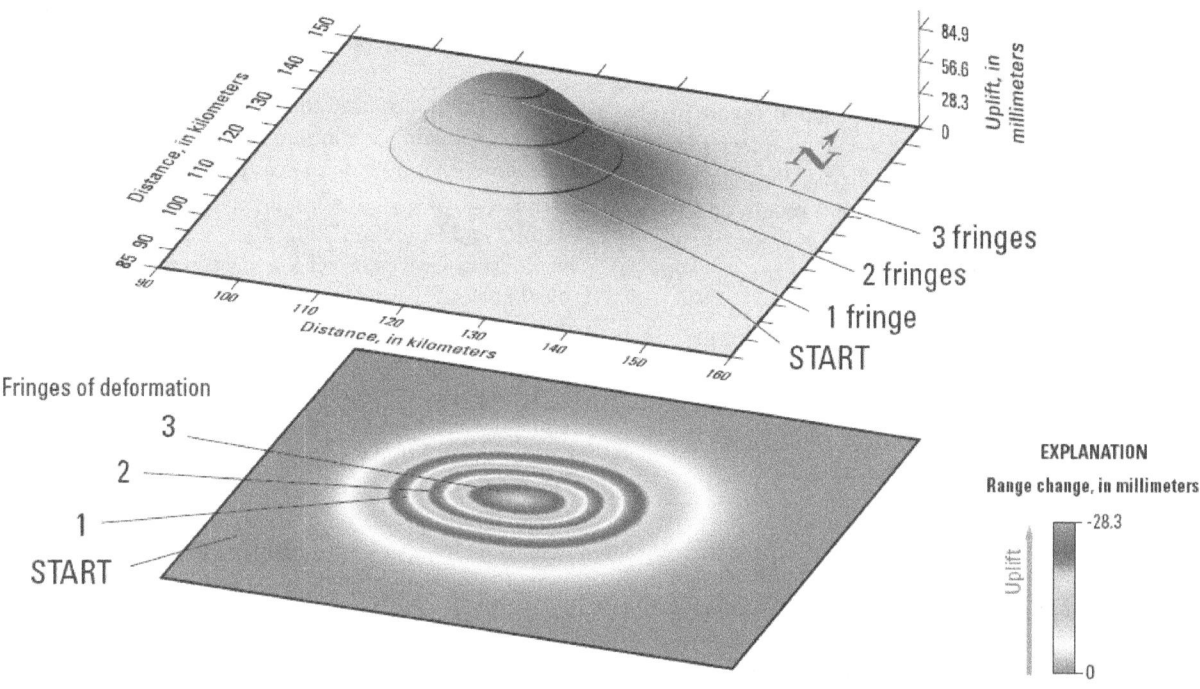

Fringes of deformation

B. Steps to interpret an interferogram

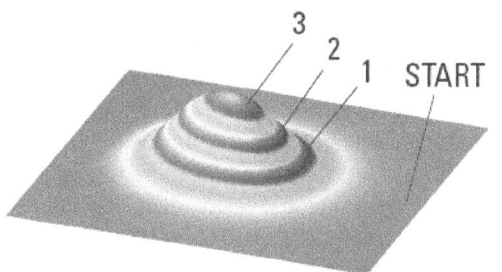

- Count the number of fringes

    From the outer red area ("START") to the center, three red fringes can be counted.

- Multiply the number of fringes by 28.3 millimeters to estimate deformation

    3 x 28.3 = 84.9 millimeters

- Determine range change direction from the scale bar

    The color sequence from the outer edge inward is red-orange-yellow-green-blue-violet, which corresponds to uplift and a decreased distance between the ground and the satellite (range).

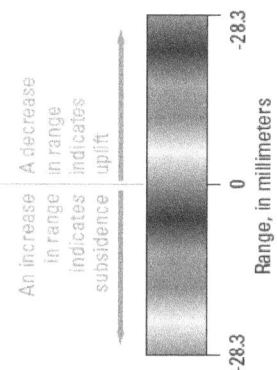

Figure 1.2.   Simplified interferogram. *A,* Approximately three fringes of deformation equaling 84.9 millimeters (mm) of uplift. *B,* Steps to interpret an interferogram.

radar scene is generated by mathematically calculating what the master SAR would look like based on orbital ephemerides and the topographic data in the DEM (fig. 1.1*D*). The topography patterns in the synthetic radar image are pixel-correlated with patterns on the SAR image such that the SAR image is precisely aligned and georeferenced to the DEM. The pixel size of the DEM determines the final resolution in the interferograms, which typically are composed of square pixels that range in edge length: 10, 30, 60, and 90 m. Interferograms can be processed at each of these pixel sizes. The smaller the pixel size the greater the computer resources required (disk space, random access memory (RAM), processing time) and the more that small changes to the land surface (tree growth, car positional changes, and so forth) influence the overall interferogram. A 30-m DEM was used as the base for the InSAR processing in this report (European Space Agency, 2007).

Removal of Topography Effects

A georeferenced-interferogram is processed to remove the effects of topography. Variations in the satellite's orbital position can create a parallax effect that is manifested in the interferograms as fringes that conform to the topography. Increased separation between the two satellite orbits results in increased sensitivity to topography and, therefore, increases the number of topographic fringes in the interferogram. Topographic fringes are removed by calculating the fringe pattern from the DEM, aligning the imagery, and then subtracting the topographic fringes from the interferogram. The greater the distance between the two satellite orbits used to create an interferogram, the more likely errors in the DEM will be seen in the imagery. Orbital positions that are within 200 m produce usable interferograms. Due to the flat topography of the Texas Gulf Coastal Plain, these topography-driven parallax effects are minor and easily identified and correlated with the topography of the Houston-Galveston region. The effects of topography were automatically removed from each interferogram as part of the standard processing approach (Schmidt and Bürgman, 2003; European Space Agency, 2007).

Unwrapping the Interferogram

Absolute distance is assigned to the interferogram through a process called unwrapping. This process converts the interferogram from the phase space (0 to 2π) to absolute distance between the ground and satellite based on sufficient data density to resolve the fringe gradient across the target area. Unwrapping refers to resolving the ambiguous phase shifts that result from phase changes that are multiples of a full (2π) cycle. There are assumptions and approaches for unwrapping the interferograms that are beyond the scope of this report and that largely do not apply to the Houston-Galveston region because of the lack of relief in land surface

and the signal coherence across the area (European Space Agency, 2007).

Limitations and Potential Considerations for Error

The techniques used to create interferograms from satellite SAR data have limitations measuring surface motion in regions with moderate to heavy vegetation, atmospheric moisture (clouds), high humidity, and high-relief topography (Schmidt and Bürgman, 2003; European Space Agency, 2007). The Houston-Galveston region has components of many of the typical noise and error sources. Each noise and error source and the approach used to mitigate its effects are described herein.

Changes in the vegetation between each satellite image are mapped as changes in the distance between ground and satellite (range). Radar signals bounce off of dense leaves, and when there is a change in the leaf canopy (summer versus winter), the range randomly changes across the vegetated land and appears in the interferograms as random color patterns, or radar incoherence, that are not associated with land-surface displacement. To mitigate the effects of changes in vegetation, interferograms were developed from radar scenes acquired close in time. Vegetation patterns change seasonally, but with data collected every 35 days, the impact of vegetation can be minimized. The Houston-Galveston region SAR scenes were selected to minimize the time period between each acquisition and study the land surface during the same time of the year, ideally during the fall and winter when there are fewer leaves. During this time period (leaf-off conditions), the radar signal penetrates the small branches and images the land surface and stable tree trunks, thus reducing incoherence. Filtering algorithms can enhance the signal quality within the interferograms and minimize the noise associated with vegetation.

The largest single source of data noise, or error, for the Houston-Galveston region was atmospheric moisture associated with clouds (fig. 1.3). Atmospheric moisture delays the radar signal as it travels from the satellite to the ground and back. This delay increases the radar signal travel time and the calculated range distance and may produce artifacts in the interferograms that equate to as much as 300 mm of delay for the most extreme conditions (fig. 1.3). Typically clouds, humidity, and fog only produce about 30–50 mm of noise in the interferograms. The error associated with atmospheric moisture was mitigated by combining or stacking radar scenes to average out any atmospheric noise and avoiding SAR scenes from days with measureable precipitation. The spatial patterns associated with clouds can be identified. For example, three radar scenes for the same ground location are examined, A, B, and C, with scene B containing clouds. The interferogram produced from scenes A and B (A–B) would show an increase in satellite range (apparent subsidence) and the interferogram produced from scenes B and C (B–C) would

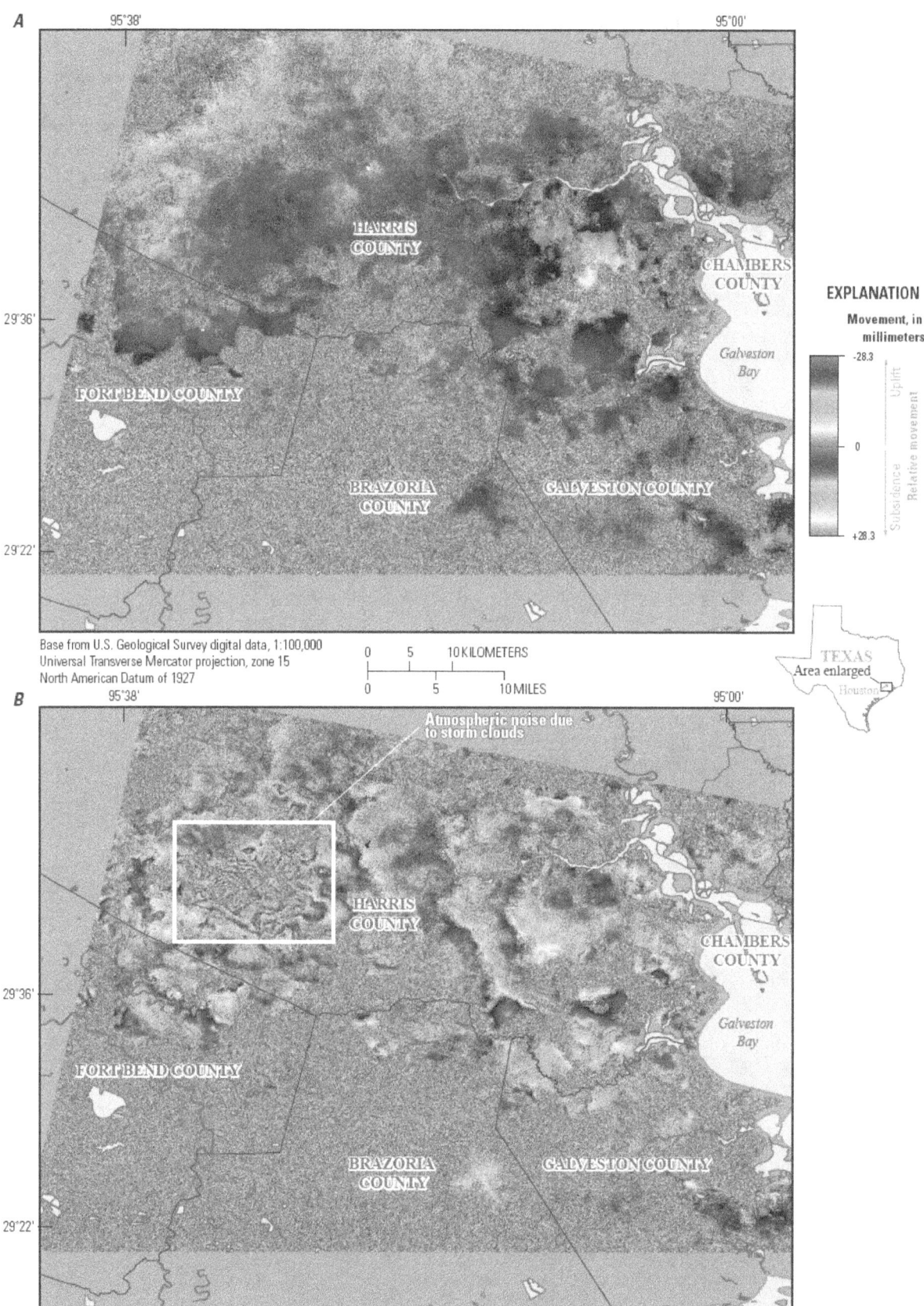

Figure 1.3. Examples of interferograms. *A*, Few clouds were present and there was minimal noise, Scene T212-F3015 for January 14, 1997, to December 30, 1997. *B*, Storm was present (box), Scene T212-F3015 for January 14, 1997, to July 13, 1999. The fringes in this interferogram are the summation of land surface motion (subsidence), noise related to the storm (box), and possible other clouds in the scene.

show a decrease in range (apparent uplift) that is opposite in magnitude of A–B. Therefore, adding A–B and B–C produces an interferogram that is free of the clouds observed in scene B. Interferograms with severe atmospheric noise were not used in the analysis. Atmospheric humidity delays the radar signal in a manner similar to atmospheric moisture, but its magnitude is smaller. To reduce error from humidity, the same techniques that were applied to atmospheric moisture section above were applied (European Space Agency, 2007).

Land surface topography is examined as part of the error assessment because errors in the DEM used to remove topography in the InSAR processing can result in topographic artifact fringes in the interferogram. To avoid topographic errors, SAR scene combinations with the least orbital variations were used. Because the Houston-Galveston region is flat, the topography errors were negligible (Schmidt and Bürgman, 2003; European Space Agency, 2007).

Appendix 2—Global Positioning System Time-series for All GPS Stations with Least-Squares Linear-Regression Lines Depicting the Directional Velocity

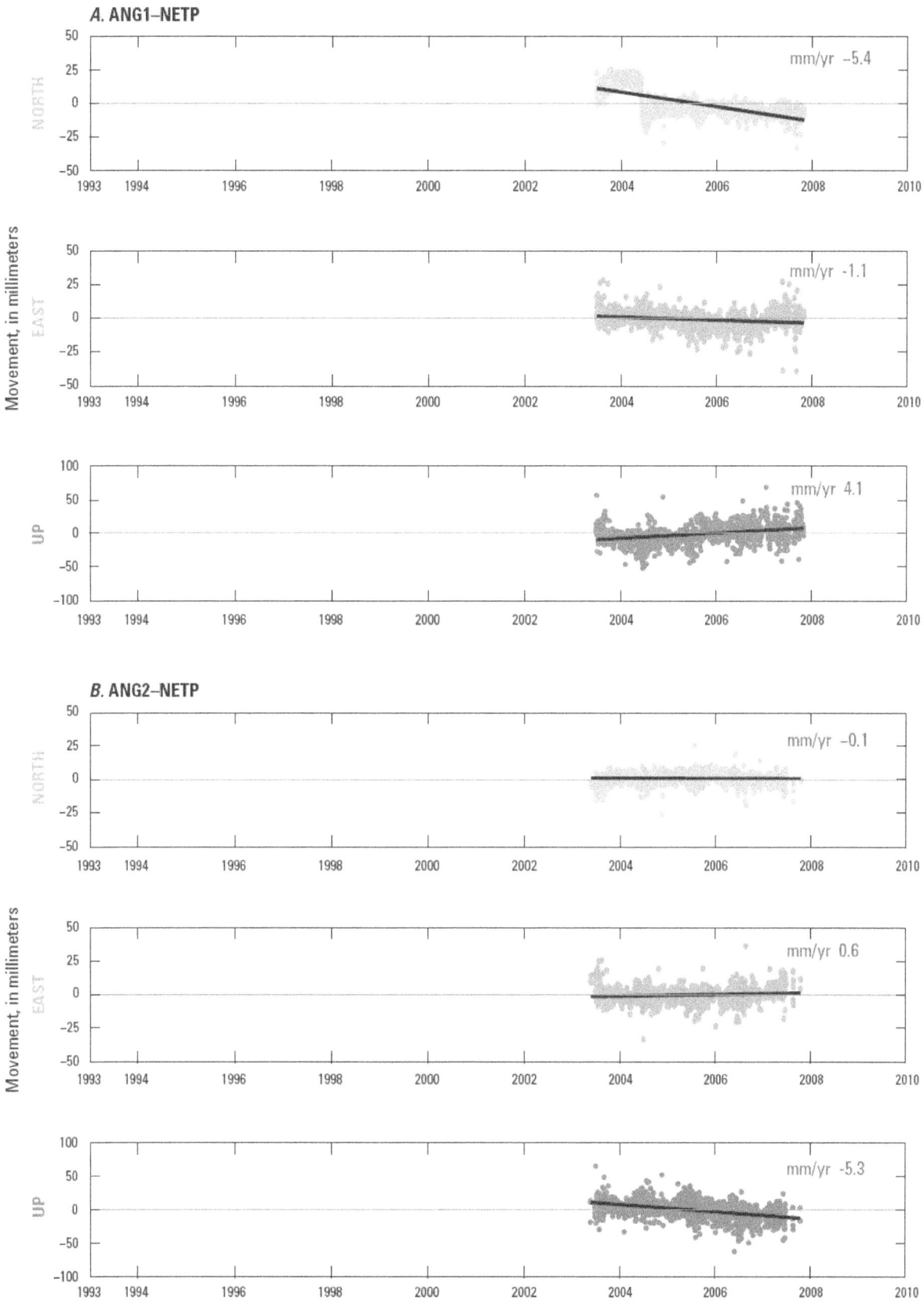

Figure 2.1. Global Positioning System (GPS) time-series of relative movement with least-squares linear-regression lines depicting the directional movement in millimeters per year (mm/yr) between *A*, Angleton 1 (ANG1) and Northeast 2250 CORS ARP (NETP) Continuously Operating Reference Station (CORS) sites and *B*, Angleton 2 (ANG2) and Northeast 2250 CORS ARP (NETP) CORS sites.

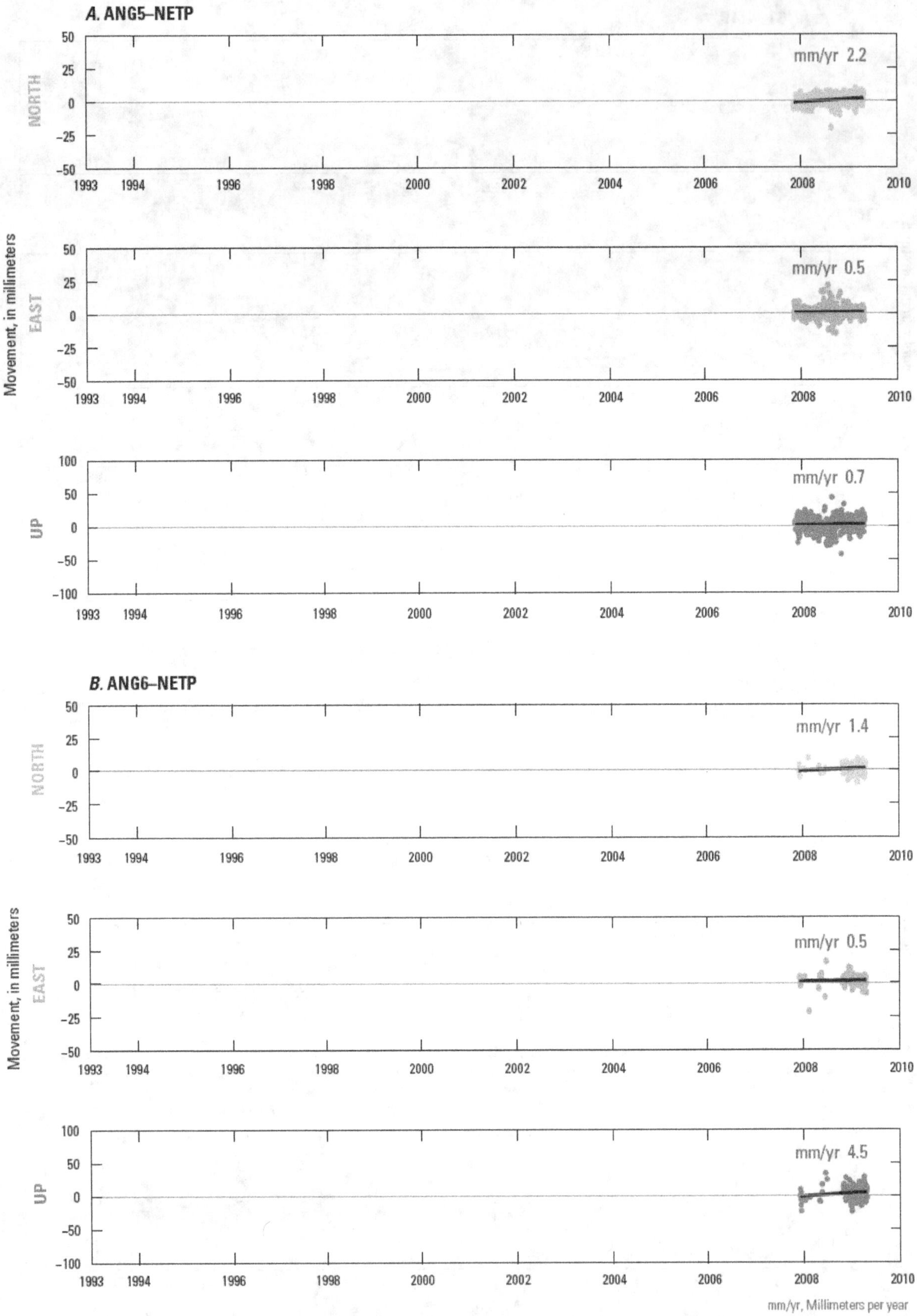

A. ANG5–NETP

B. ANG6–NETP

mm/yr, Millimeters per year

Figure 2.2.   Global Positioning System (GPS) time-series of relative movement with least-squares linear-regression lines depicting the directional movement in millimeters per year (mm/yr) between *A*, Angleton 5 (ANG5) and Northeast 2250 CORS ARP (NETP) Continuously Operating Reference Station (CORS) sites and *B*, Angleton 6 (ANG6) and Northeast 2250 CORS ARP (NETP) CORS sites.

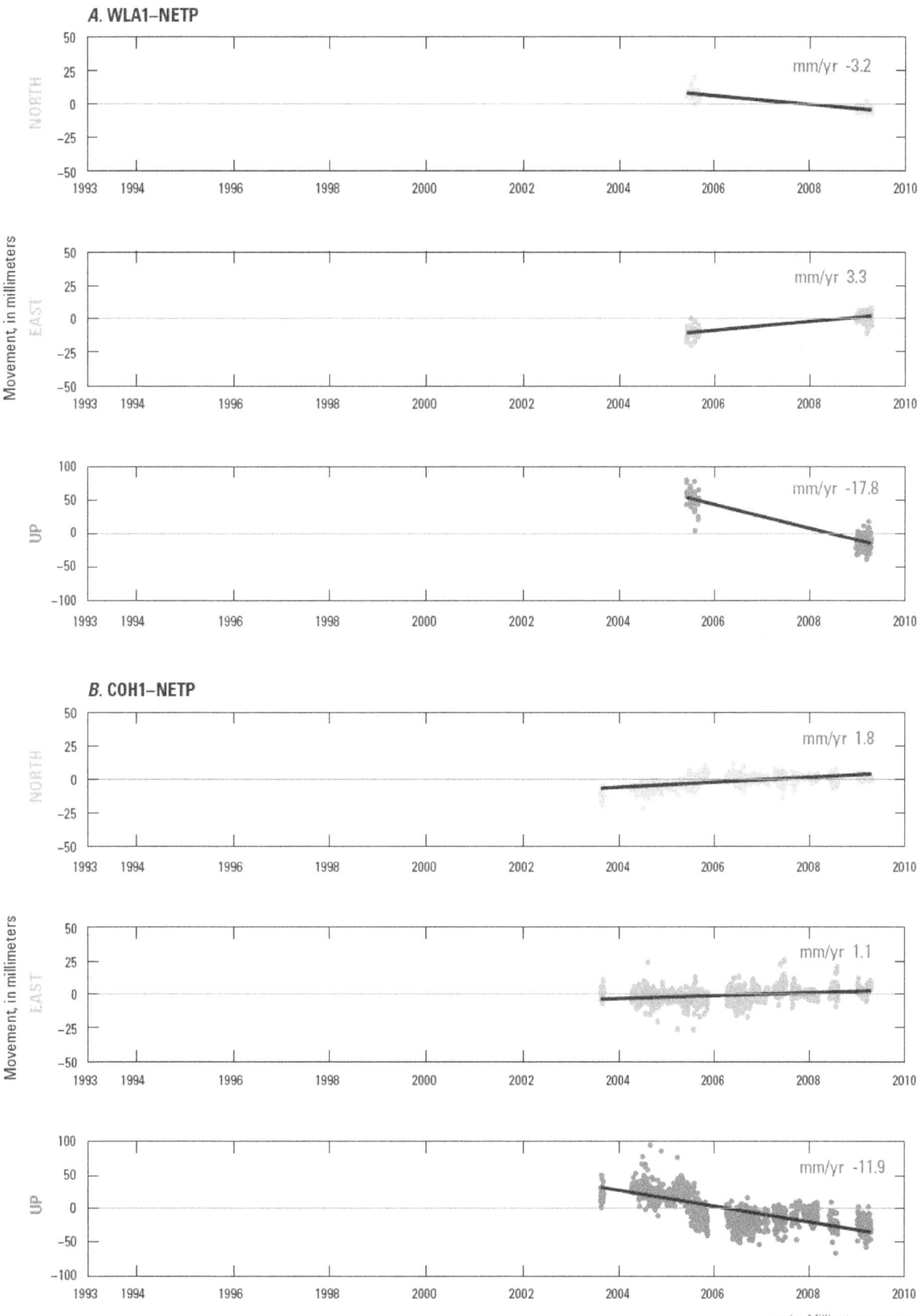

Figure 2.3.   Global Positioning System (GPS) time-series of relative movement with least-squares linear-regression lines depicting the directional movement in millimeters per year (mm/yr) between *A*, Tomball COOP (WLA1) and Northeast 2250 CORS ARP (NETP) Continuously Operating Reference Station (CORS) sites and *B*, C OF Houston COOP (COH1) and Northeast 2250 CORS ARP (NETP) CORS sites.

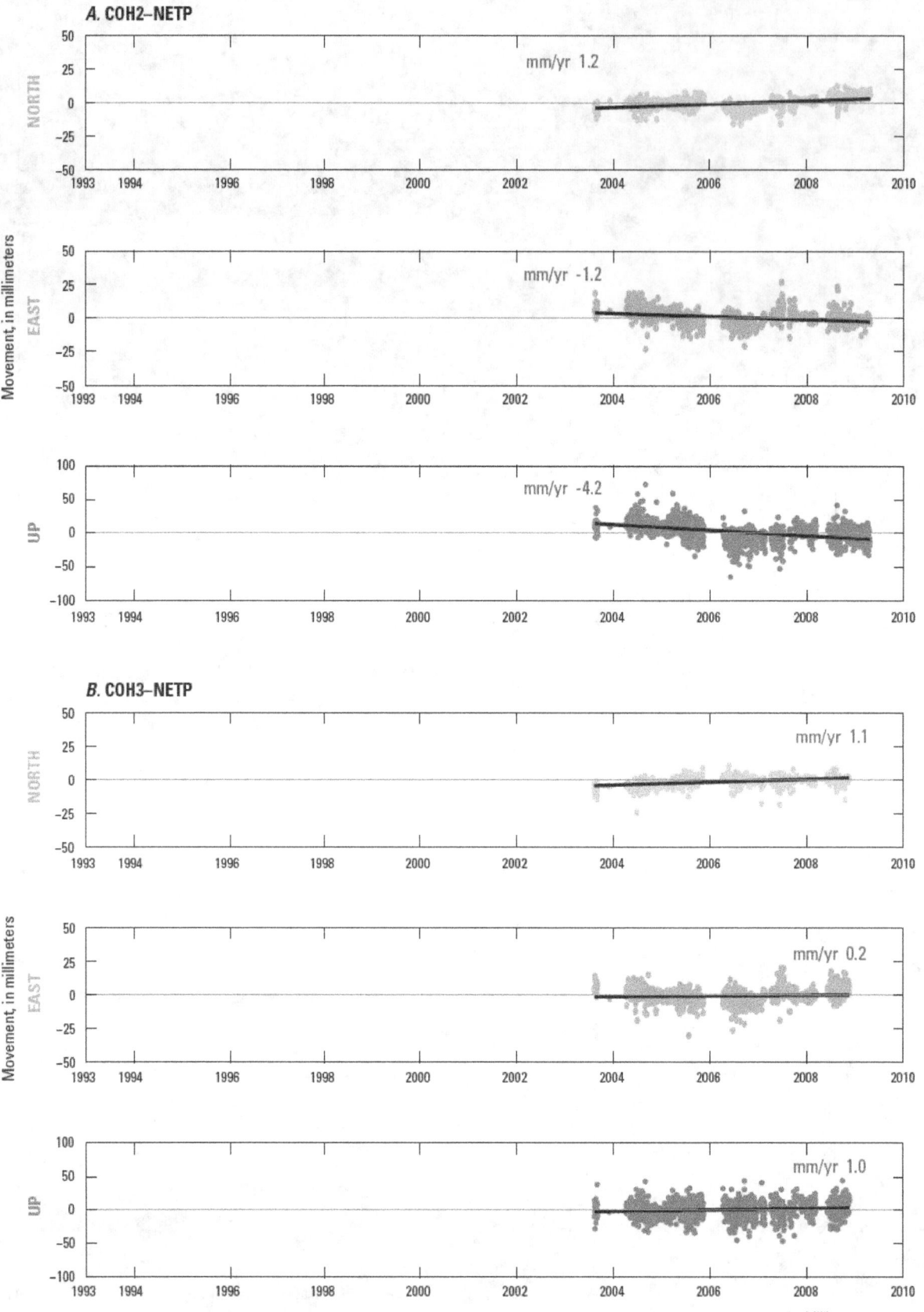

Figure 2.4. Global Positioning System (GPS) time-series of relative movement with least-squares linear-regression lines depicting the directional movement in millimeters per year (mm/yr) between *A*, Houston 2 COOP (COH2) and Northeast 2250 CORS ARP (NETP) Continuously Operating Reference Station (CORS) sites and *B*, Houston 3 COOP (COH3) and Northeast 2250 CORS ARP (NETP) CORS sites.

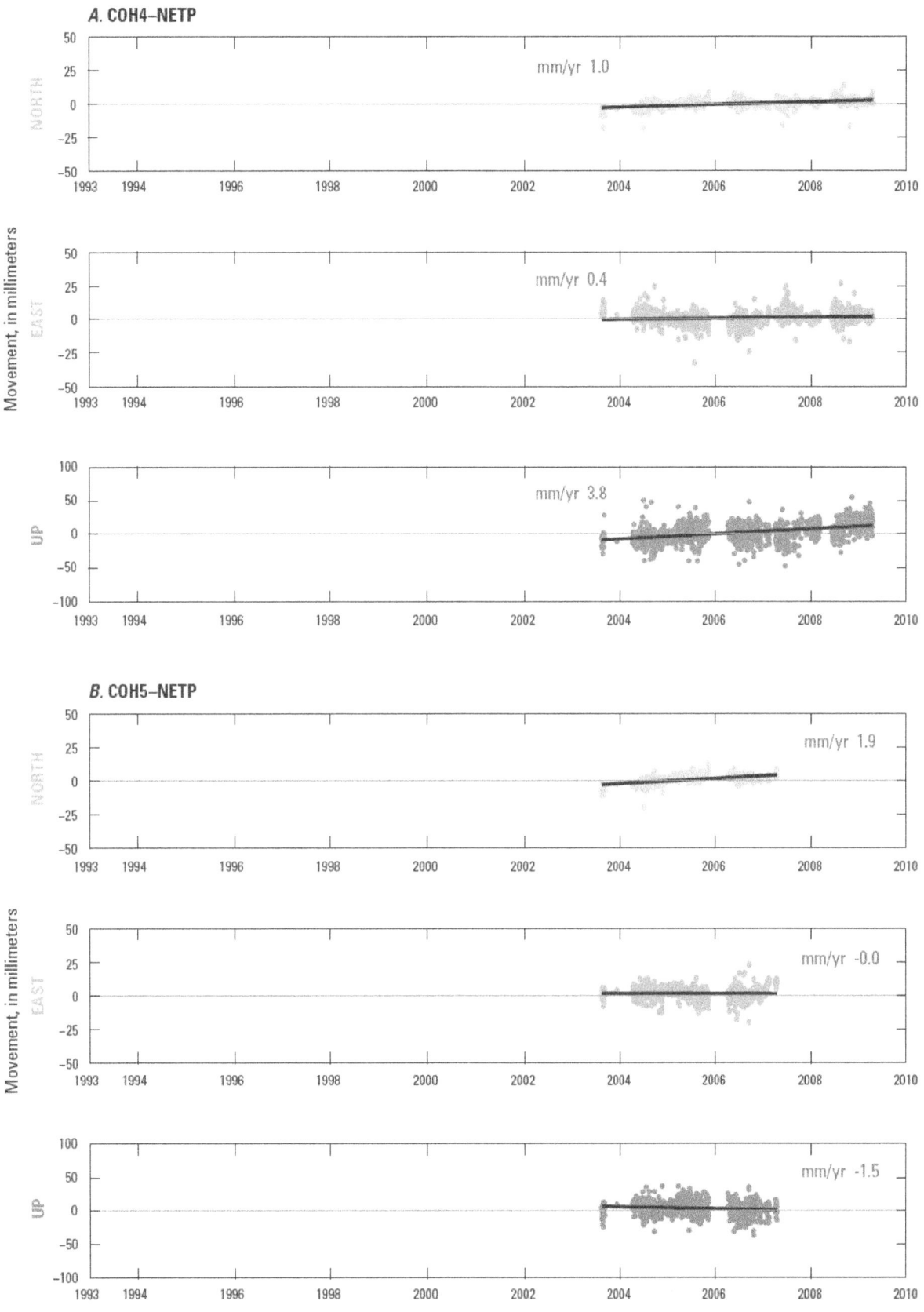

Figure 2.5. Global Positioning System (GPS) time-series of relative movement with least-squares linear-regression lines depicting the directional movement in millimeters per year (mm/yr) between *A*, Houston 4 COOP (COH4) and Northeast 2250 CORS ARP (NETP) Continuously Operating Reference Station (CORS) sites and *B*, Houston 5 COOP (COH5) and Northeast 2250 CORS ARP (NETP) CORS sites.

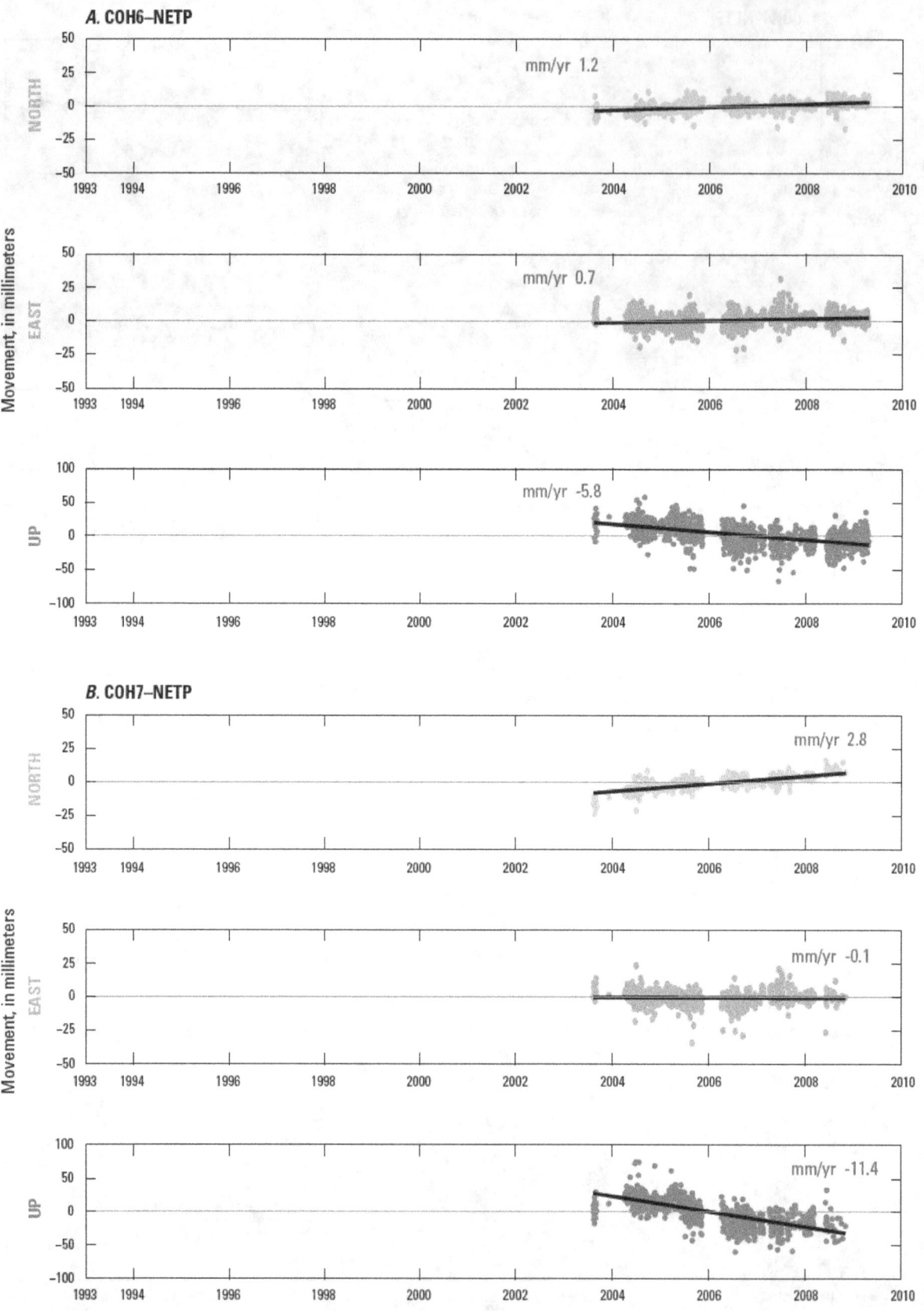

Figure 2.6.   Global Positioning System (GPS) time-series of relative movement with least-squares linear-regression lines depicting the directional movement in millimeters per year (mm/yr) between *A,* Houston 6 COOP (COH6) and Northeast 2250 CORS ARP (NETP) Continuously Operating Reference Station (CORS) sites and *B,* Houston 7 COOP (COH7) and Northeast 2250 CORS ARP (NETP) CORS sites.

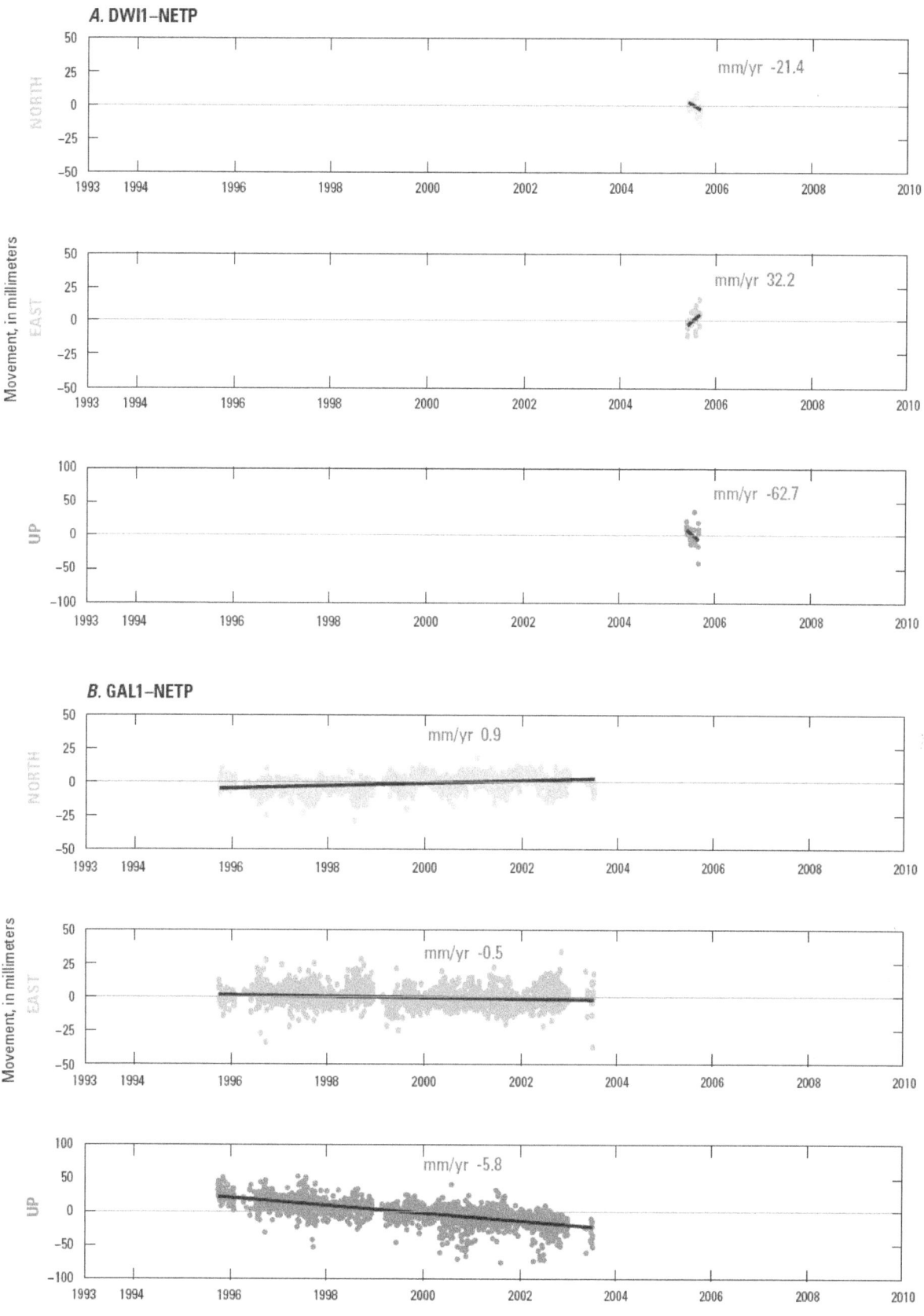

Figure 2.7.   Global Positioning System (GPS) time-series of relative movement with least-squares linear-regression lines depicting the directional movement in millimeters per year (mm/yr) between *A*, Clute COOP (DWI1) and Northeast 2250 CORS ARP (NETP) Continuously Operating Reference Station (CORS) sites and *B*, Galveston 1 (GAL1) and Northeast 2250 CORS ARP (NETP) CORS sites.

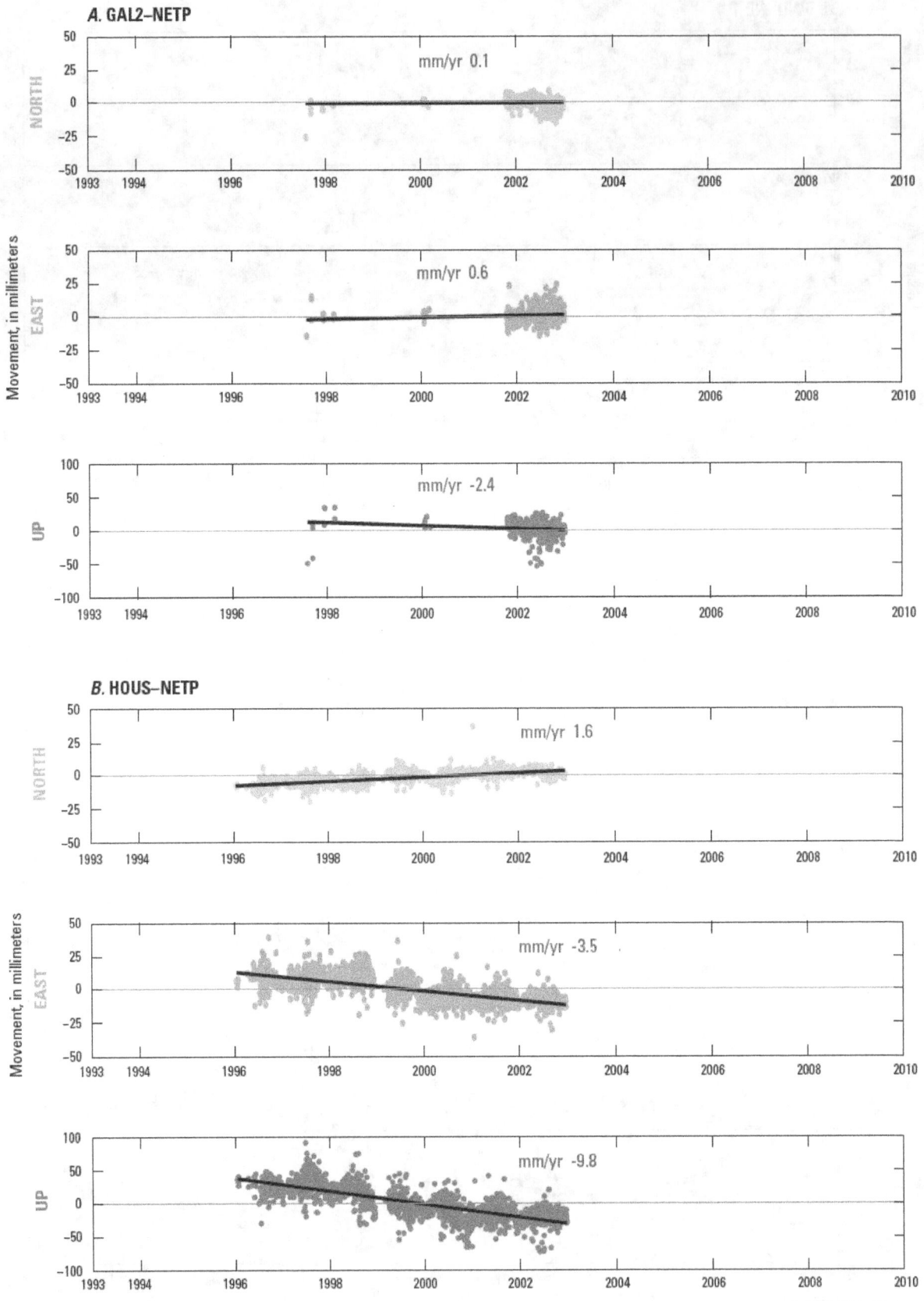

Figure 2.8. Global Positioning System (GPS) time-series of relative movement with least-squares linear-regression lines depicting the directional movement in millimeters per year (mm/yr) between *A*, Galveston 2 (GAL2) and Northeast 2250 CORS ARP (NETP) Continuously Operating Reference Station (CORS) sites and *B*, Houston (HOUS) and Northeast 2250 CORS ARP (NETP) CORS sites.

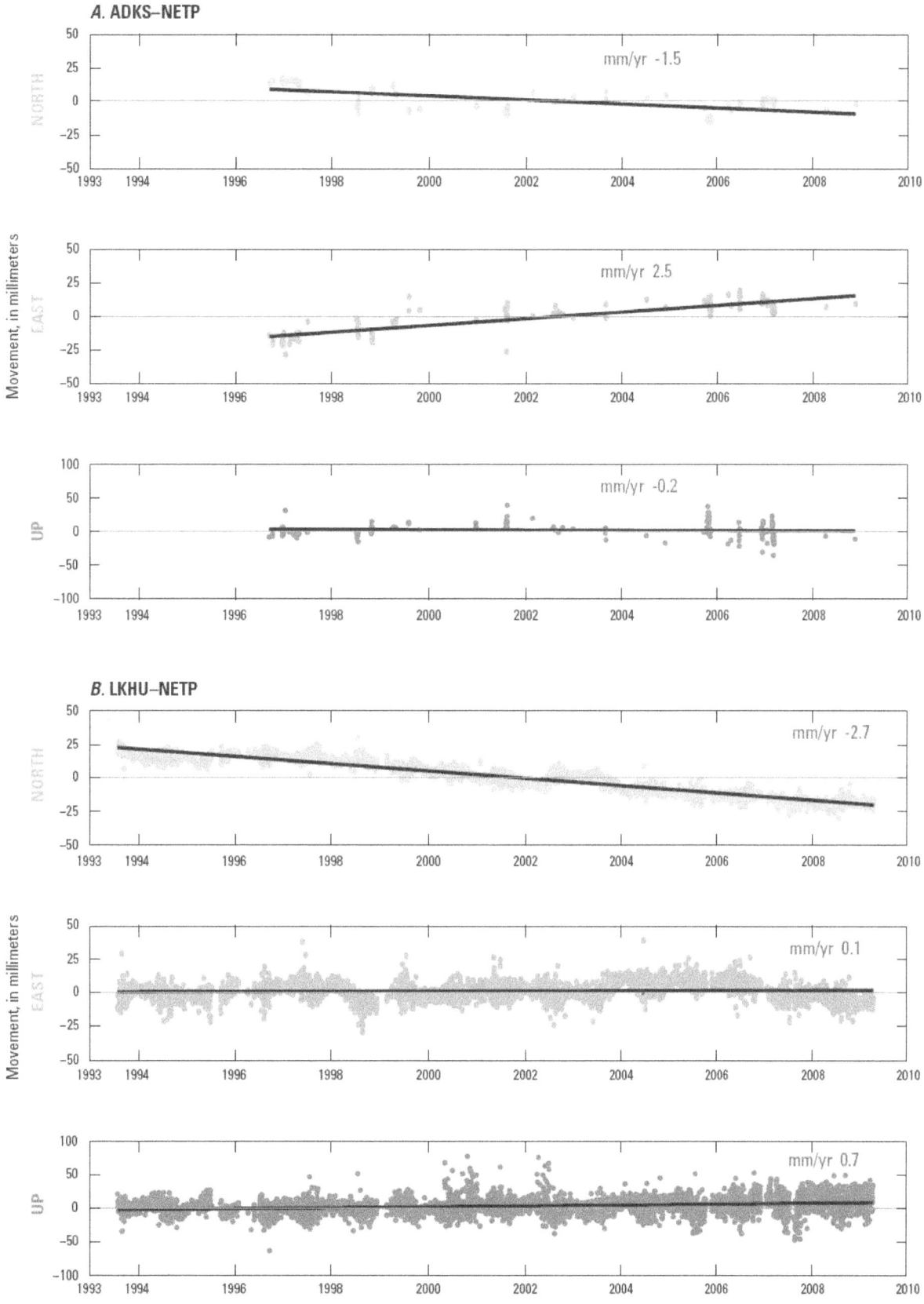

Figure 2.9. Global Positioning System (GPS) time-series of relative movement with least-squares linear-regression lines depicting the directional movement in millimeters per year (mm/yr) between *A*, Addicks 1795 CORS ARP (ADKS) and Northeast 2250 CORS ARP (NETP) Continuously Operating Reference Station (CORS) sites and *B*, Lake Houston (LKHU) and Northeast 2250 CORS ARP (NETP) CORS sites.

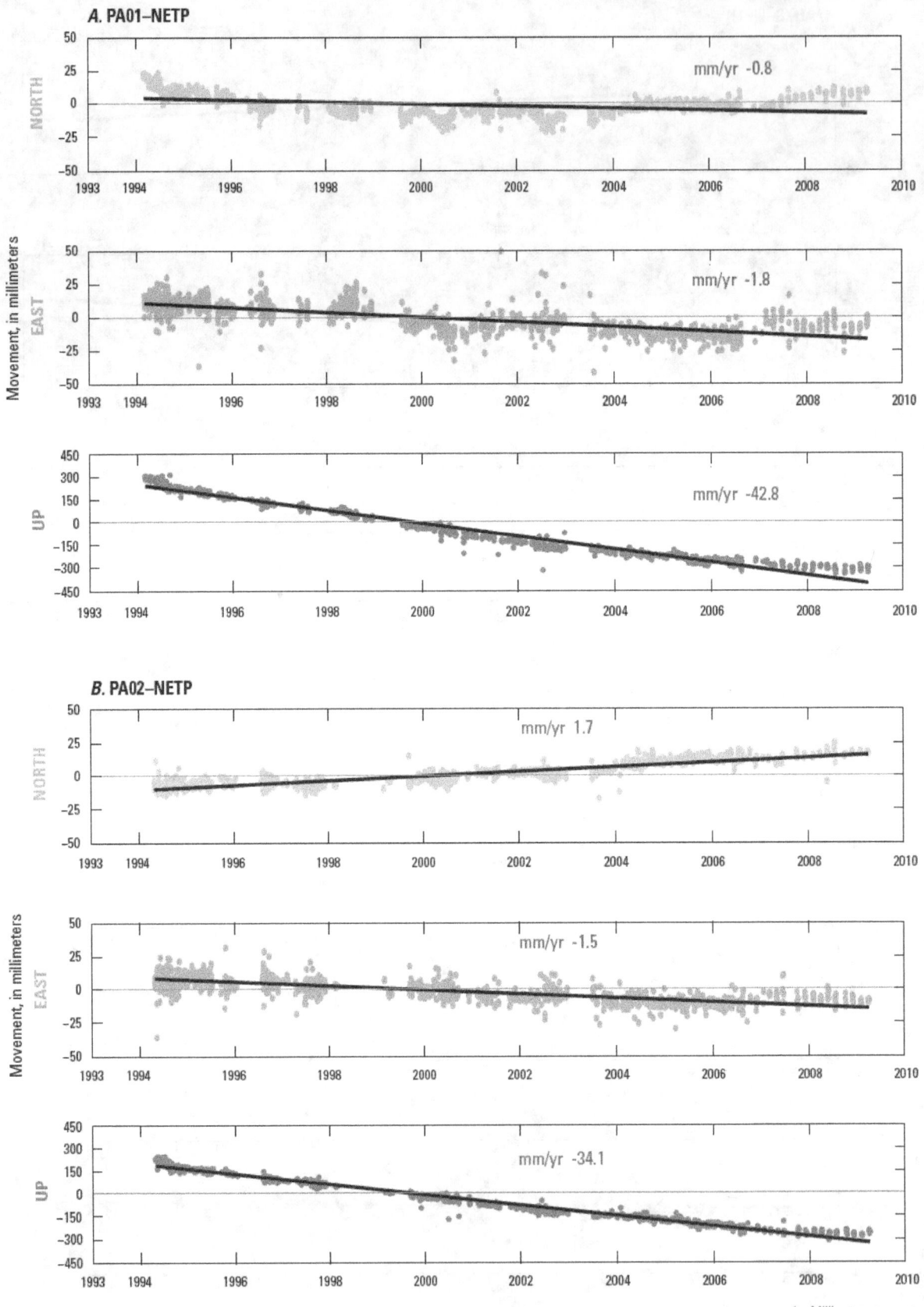

Figure 2.10.   Global Positioning System (GPS) time-series of relative movement with least-squares linear-regression lines depicting the directional movement in millimeters per year (mm/yr) between *A*, PAM 01 (PA01) and Northeast 2250 CORS ARP (NETP) Continuously Operating Reference Station (CORS) sites and *B*, PAM 02 (PA02) and Northeast 2250 CORS ARP (NETP) CORS sites.

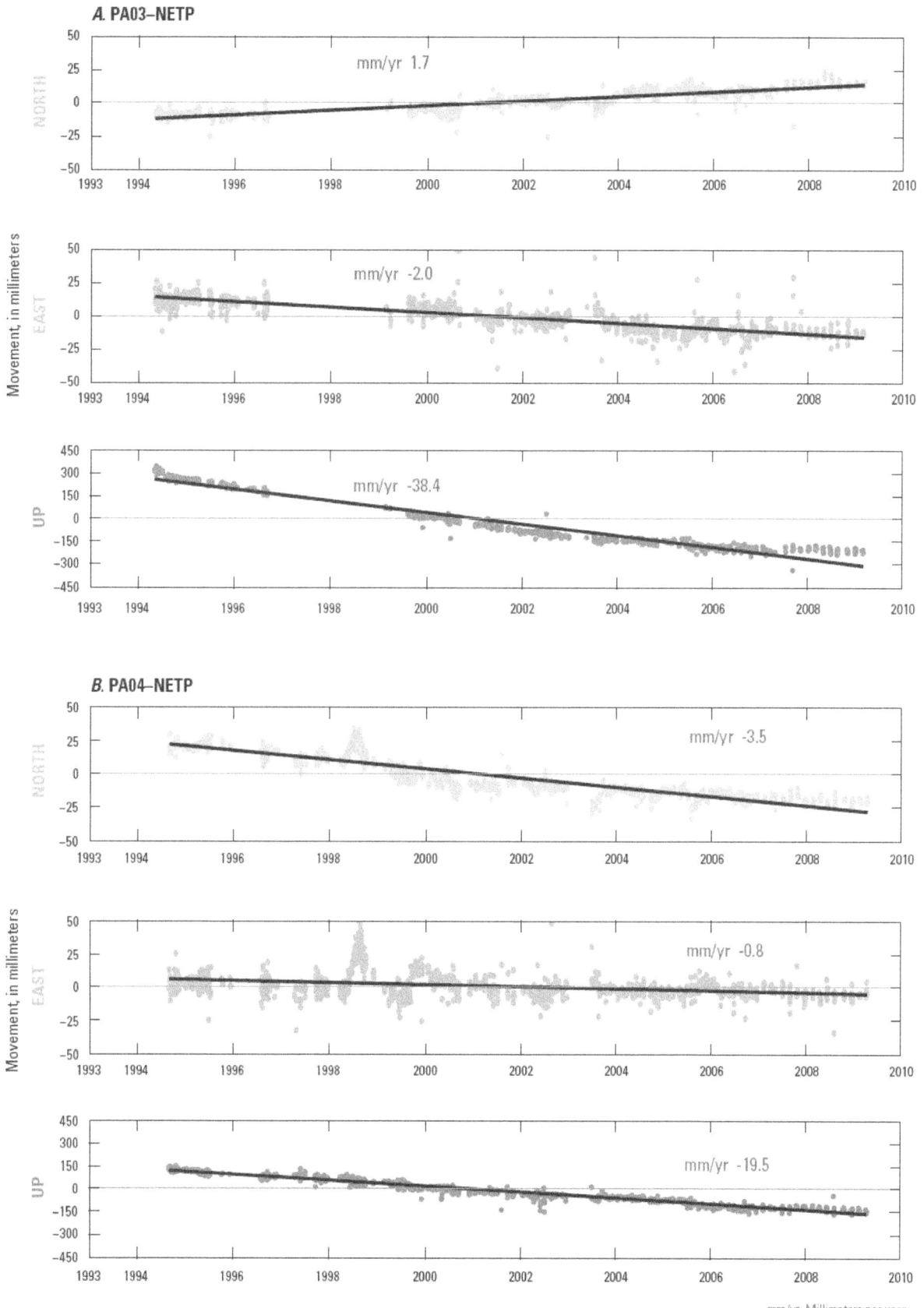

Figure 2.11.   Global Positioning System (GPS) time-series of relative movement with least-squares linear-regression lines depicting the directional movement in millimeters per year (mm/yr) between *A*, PAM 03 (PA03) and Northeast 2250 CORS ARP (NETP) Continuously Operating Reference Station (CORS) sites and *B*, PAM 04 (PA04) and Northeast 2250 CORS ARP (NETP) CORS sites.

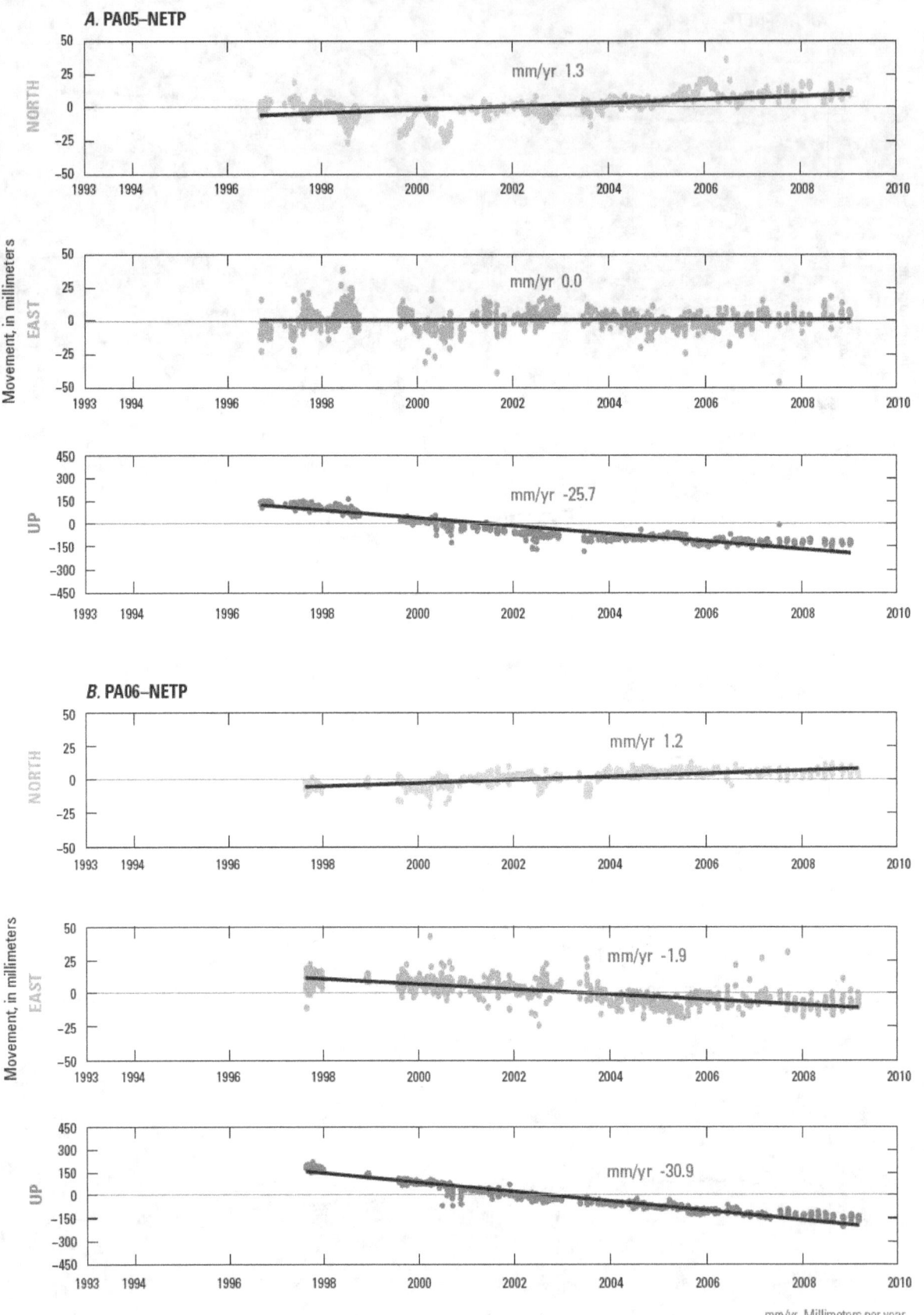

A. PA05–NETP

B. PA06–NETP

mm/yr, Millimeters per year

Figure 2.12.   Global Positioning System (GPS) time-series of relative movement with least-squares linear-regression lines depicting the directional movement in millimeters per year (mm/yr) between *A*, PAM 05 (PA05) and Northeast 2250 CORS ARP (NETP) Continuously Operating Reference Station (CORS) sites and *B*, PAM 06 (PA06) and Northeast 2250 CORS ARP (NETP) CORS sites.

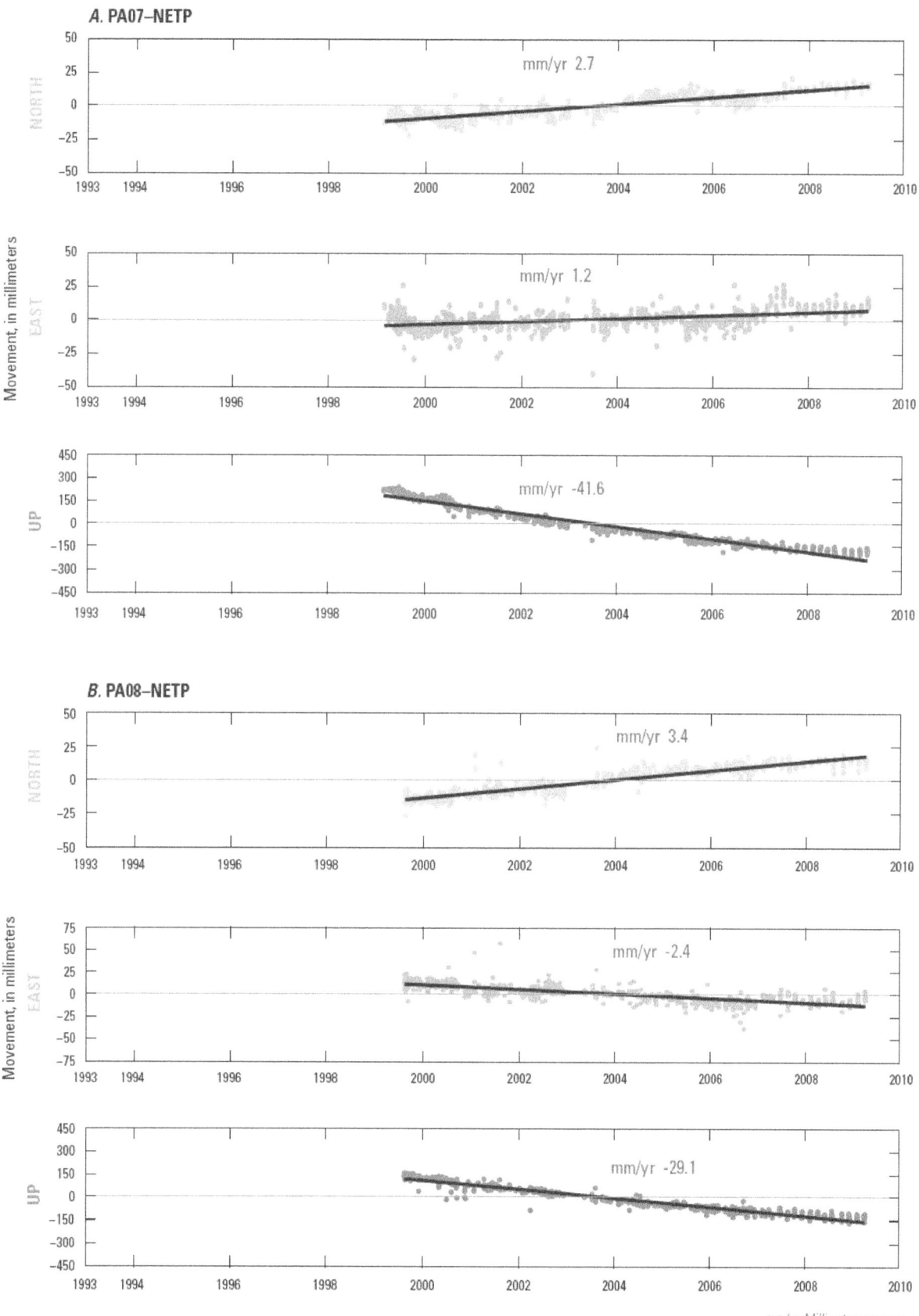

Figure 2.13. Global Positioning System (GPS) time-series of relative movement with least-squares linear-regression lines depicting the directional movement in millimeters per year (mm/yr) between *A*, PAM 07 (PA07) and Northeast 2250 CORS ARP (NETP) Continuously Operating Reference Station (CORS) sites and *B*, PAM 08 (PA08) and Northeast 2250 CORS ARP (NETP) CORS sites.

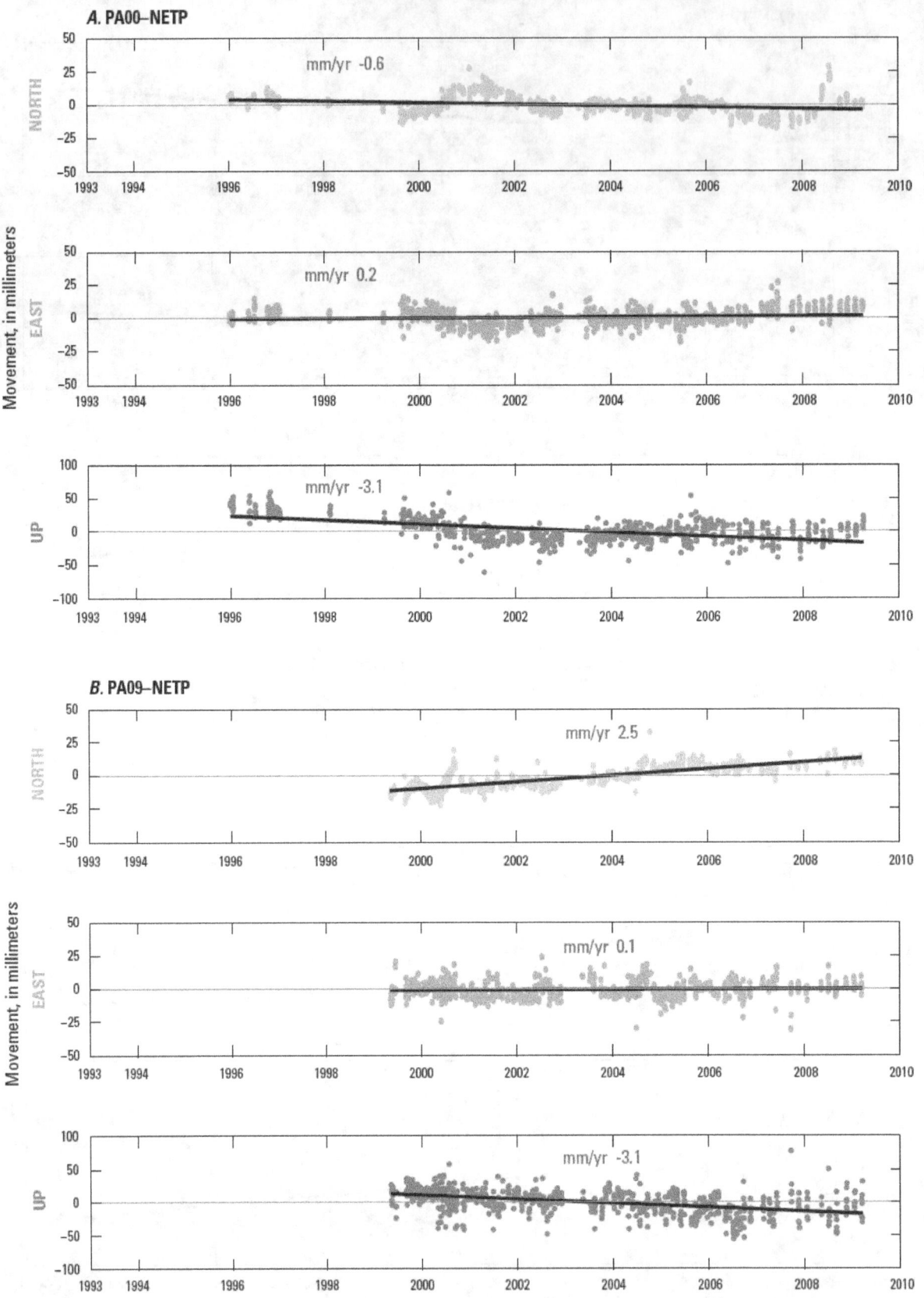

Figure 2.14.   Global Positioning System (GPS) time-series of relative movement with least-squares linear-regression lines depicting the directional movement in millimeters per year (mm/yr) between *A*, PAM 00 (PA00) and Northeast 2250 CORS ARP (NETP) Continuously Operating Reference Station (CORS) sites and *B*, PAM 09 (PA09) and Northeast 2250 CORS ARP (NETP) CORS sites.

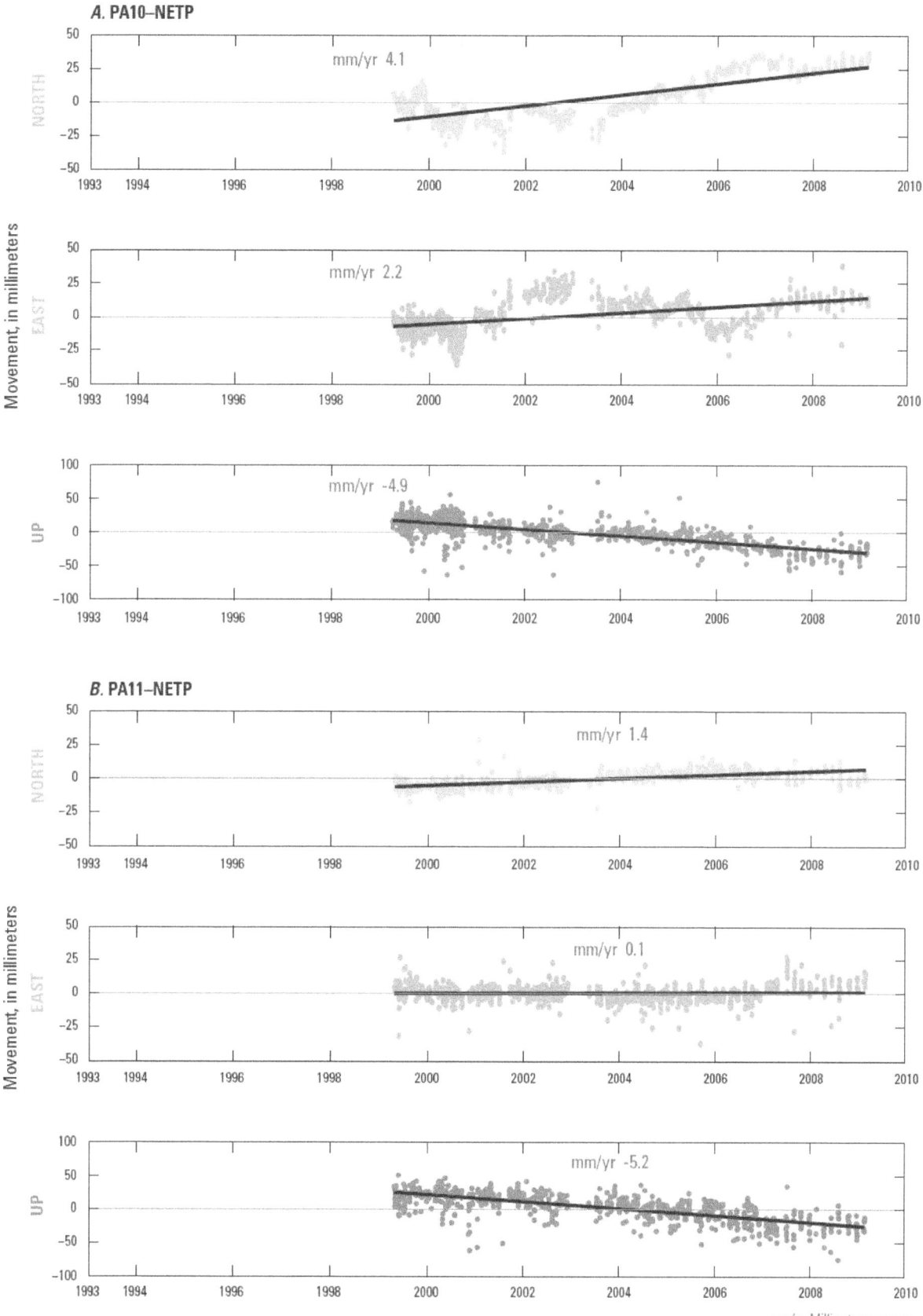

Figure 2.15. Global Positioning System (GPS) time-series of relative movement with least-squares linear-regression lines depicting the directional movement in millimeters per year (mm/yr) between *A*, PAM 10 (PA10) and Northeast 2250 CORS ARP (NETP) Continuously Operating Reference Station (CORS) sites and *B*, PAM 11 (PA11) and Northeast 2250 CORS ARP (NETP) CORS sites.

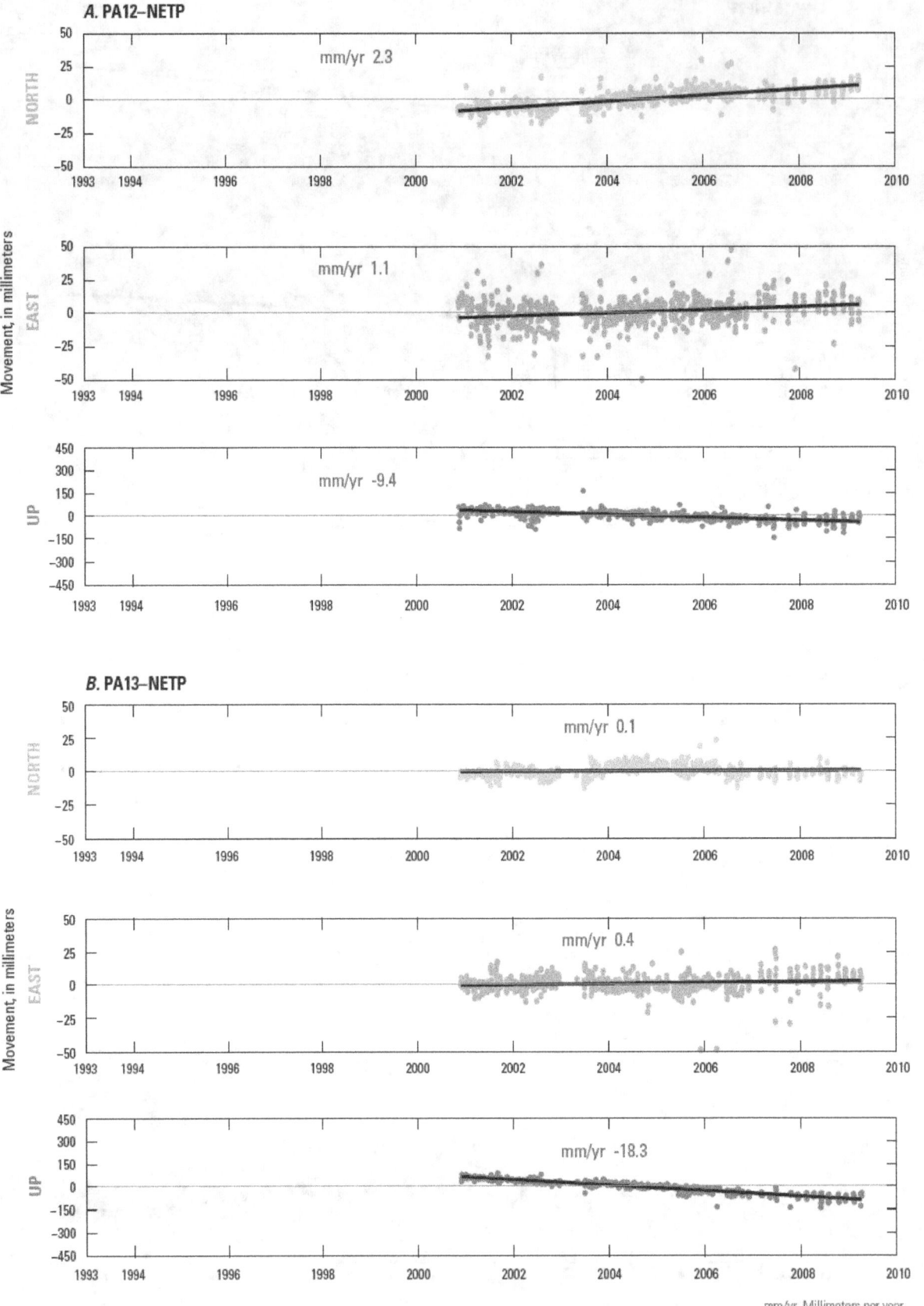

Figure 2.16.   Global Positioning System (GPS) time-series of relative movement with least-squares linear-regression lines depicting the directional movement in millimeters per year (mm/yr) between A, PAM 12 (PA12) and Northeast 2250 CORS ARP (NETP) Continuously Operating Reference Station (CORS) sites and B, PAM 13 (PA13) and Northeast 2250 CORS ARP (NETP) CORS sites.

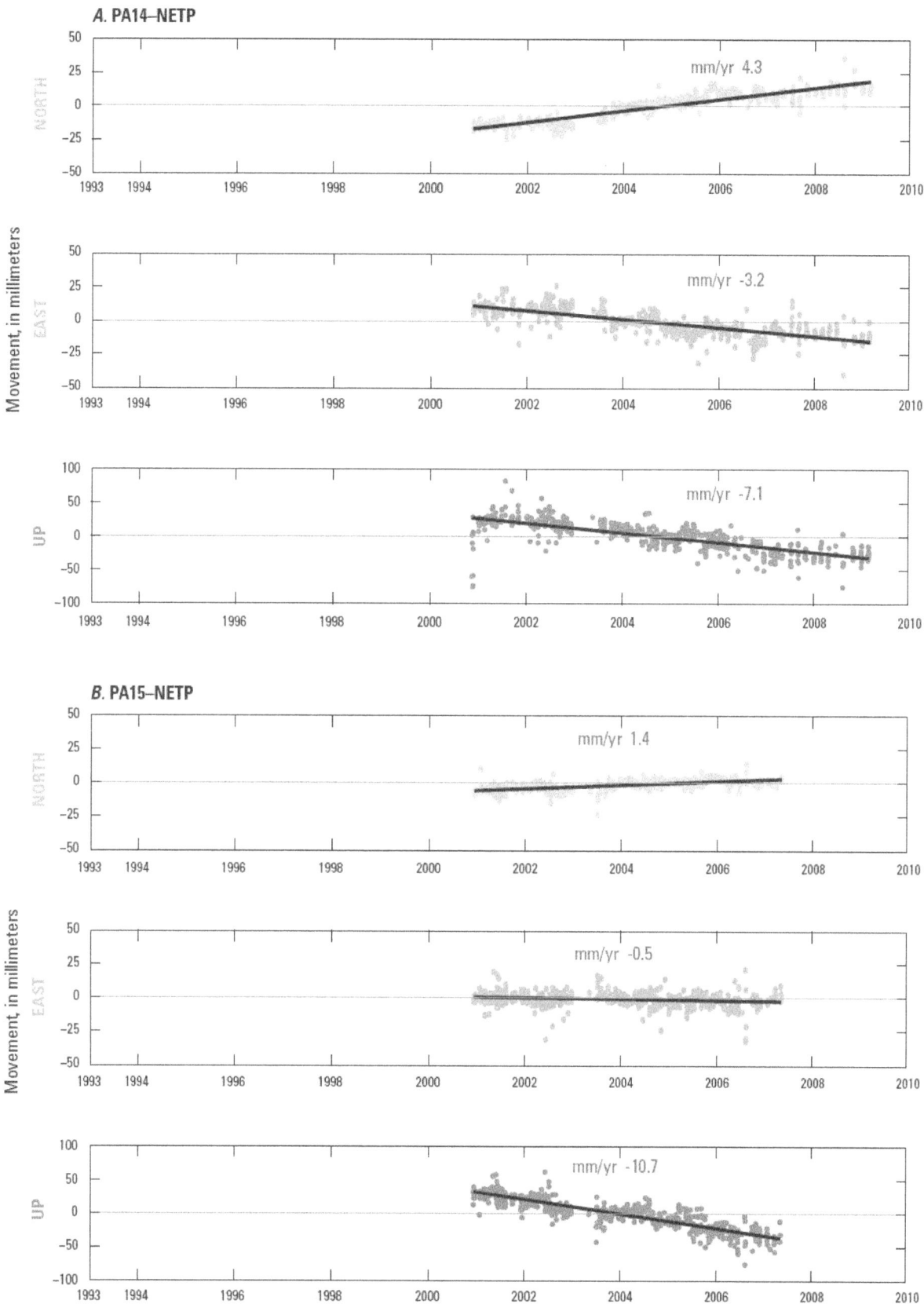

Figure 2.17.   Global Positioning System (GPS) time-series of relative movement with least-squares linear-regression lines depicting the directional movement in millimeters per year (mm/yr) between *A*, PAM 14 (PA14) and Northeast 2250 CORS ARP (NETP) Continuously Operating Reference Station (CORS) sites and *B*, PAM 15 (PA15) and Northeast 2250 CORS ARP (NETP) CORS sites.

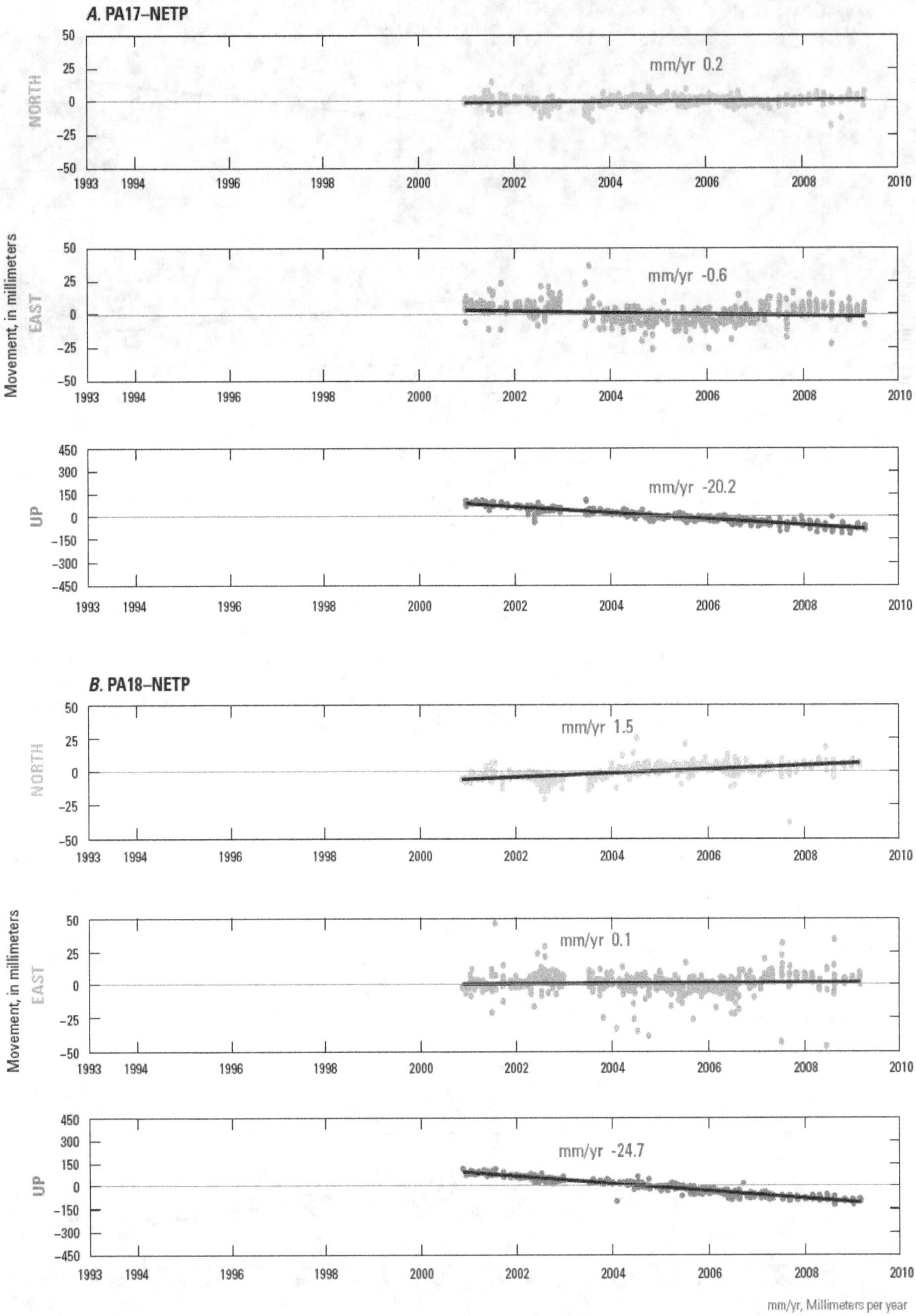

A. PA17–NETP

B. PA18–NETP

mm/yr, Millimeters per year

Figure 2.18.   Global Positioning System (GPS) time-series of relative movement with least-squares linear-regression lines depicting the directional movement in millimeters per year (mm/yr) between *A*, PAM 17 (PA17) and Northeast 2250 CORS ARP (NETP) Continuously Operating Reference Station (CORS) sites and *B*, PAM 18 (PA18) and Northeast 2250 CORS ARP (NETP) CORS sites.

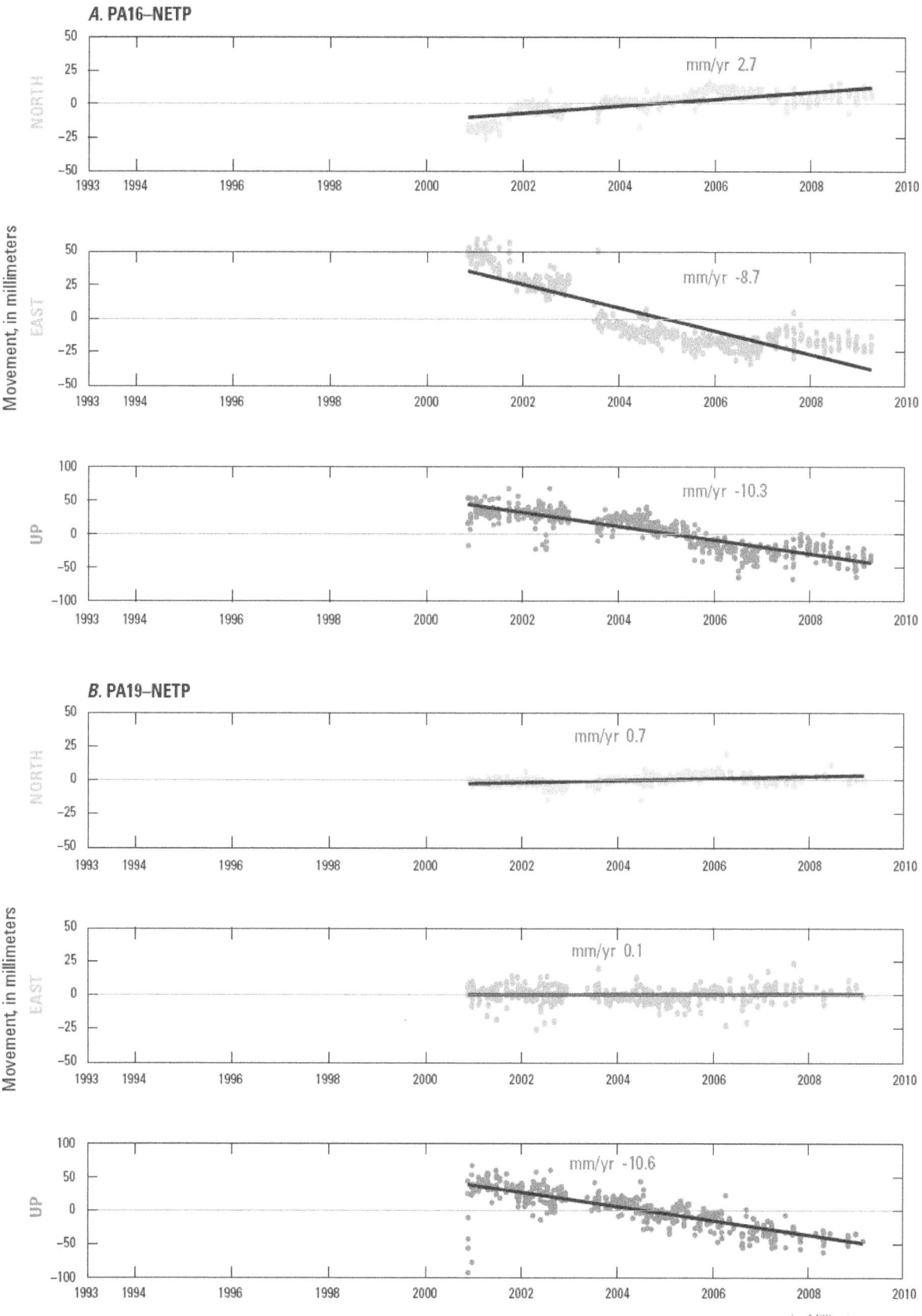

Figure 2.19.   Global Positioning System (GPS) time-series of relative movement with least-squares linear-regression lines depicting the directional movement in millimeters per year (mm/yr) between A, PAM 16 (PA16) and Northeast 2250 CORS ARP (NETP) Continuously Operating Reference Station (CORS) sites and B, PAM 19 (PA19) and Northeast 2250 CORS ARP (NETP) CORS sites.

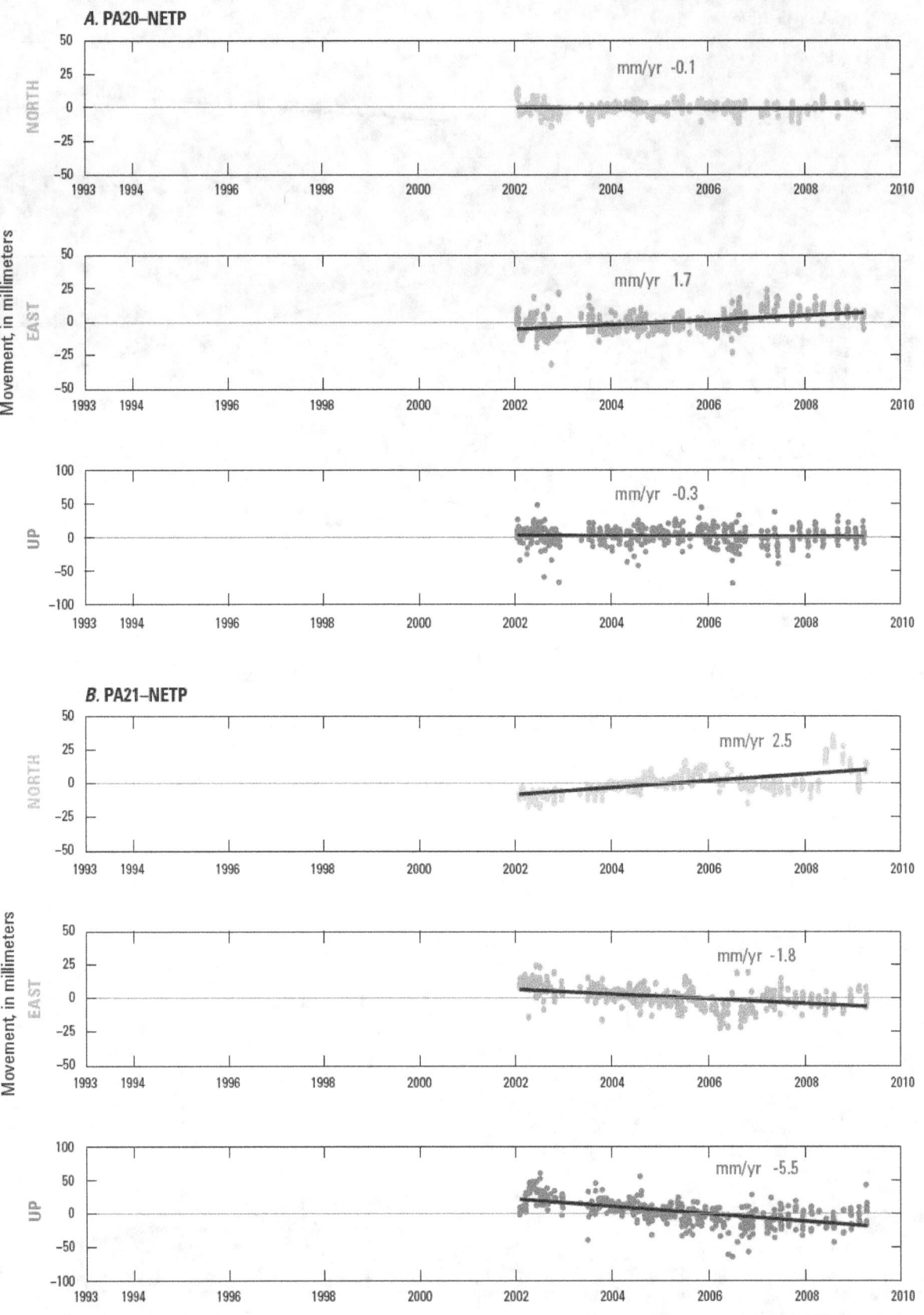

Figure 2.20. Global Positioning System (GPS) time-series of relative movement with least-squares linear-regression lines depicting the directional movement in millimeters per year (mm/yr) between *A*, PAM 20 (PA20) and Northeast 2250 CORS ARP (NETP) Continuously Operating Reference Station (CORS) sites and *B*, PAM 21 (PA21) and Northeast 2250 CORS ARP (NETP) CORS sites.

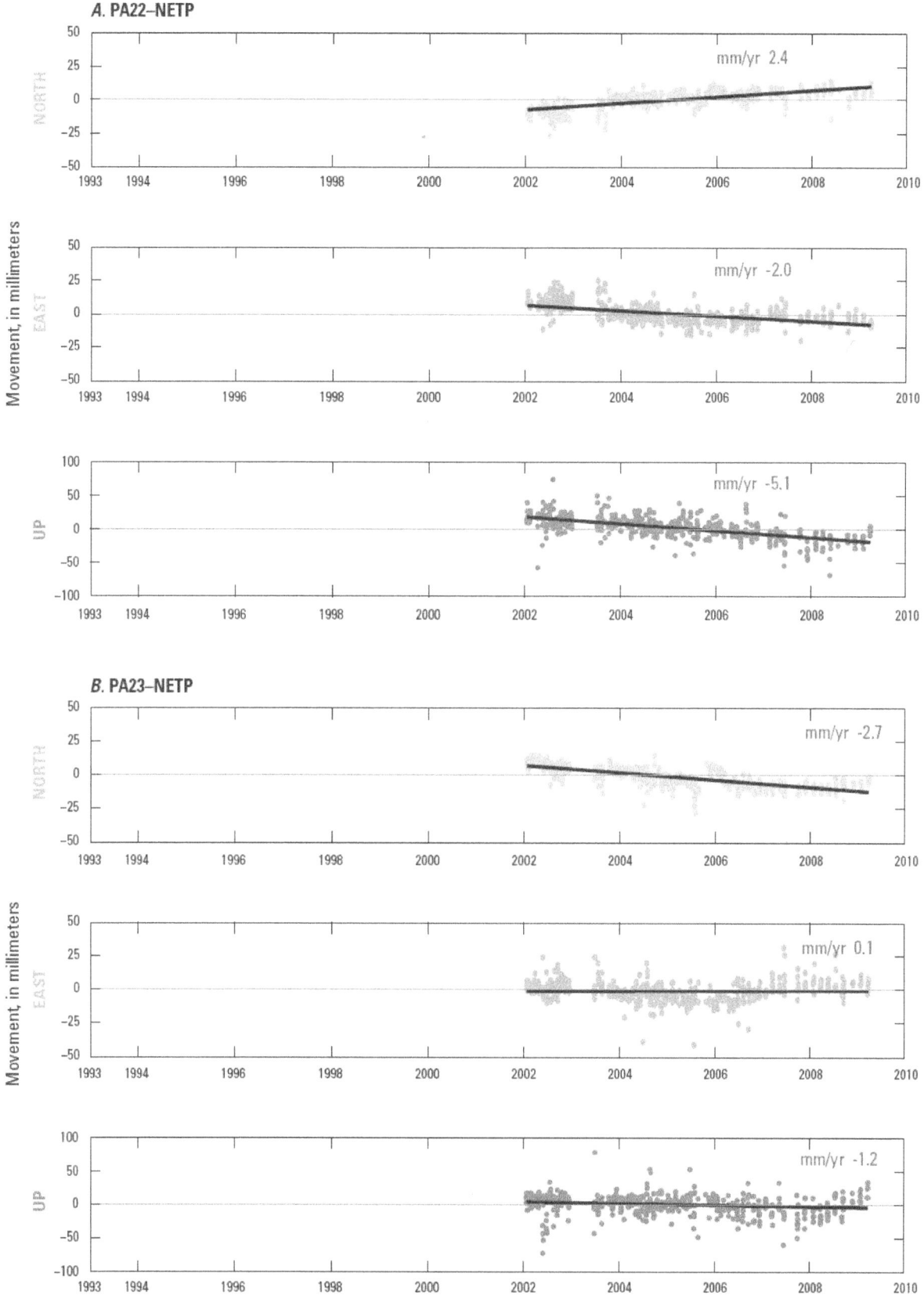

Figure 2.21.   Global Positioning System (GPS) time-series of relative movement with least-squares linear-regression lines depicting the directional movement in millimeters per year (mm/yr) between *A*, PAM 22 (PA22) and Northeast 2250 CORS ARP (NETP) Continuously Operating Reference Station (CORS) sites and *B*, PAM 23 (PA23) and Northeast 2250 CORS ARP (NETP) CORS sites.

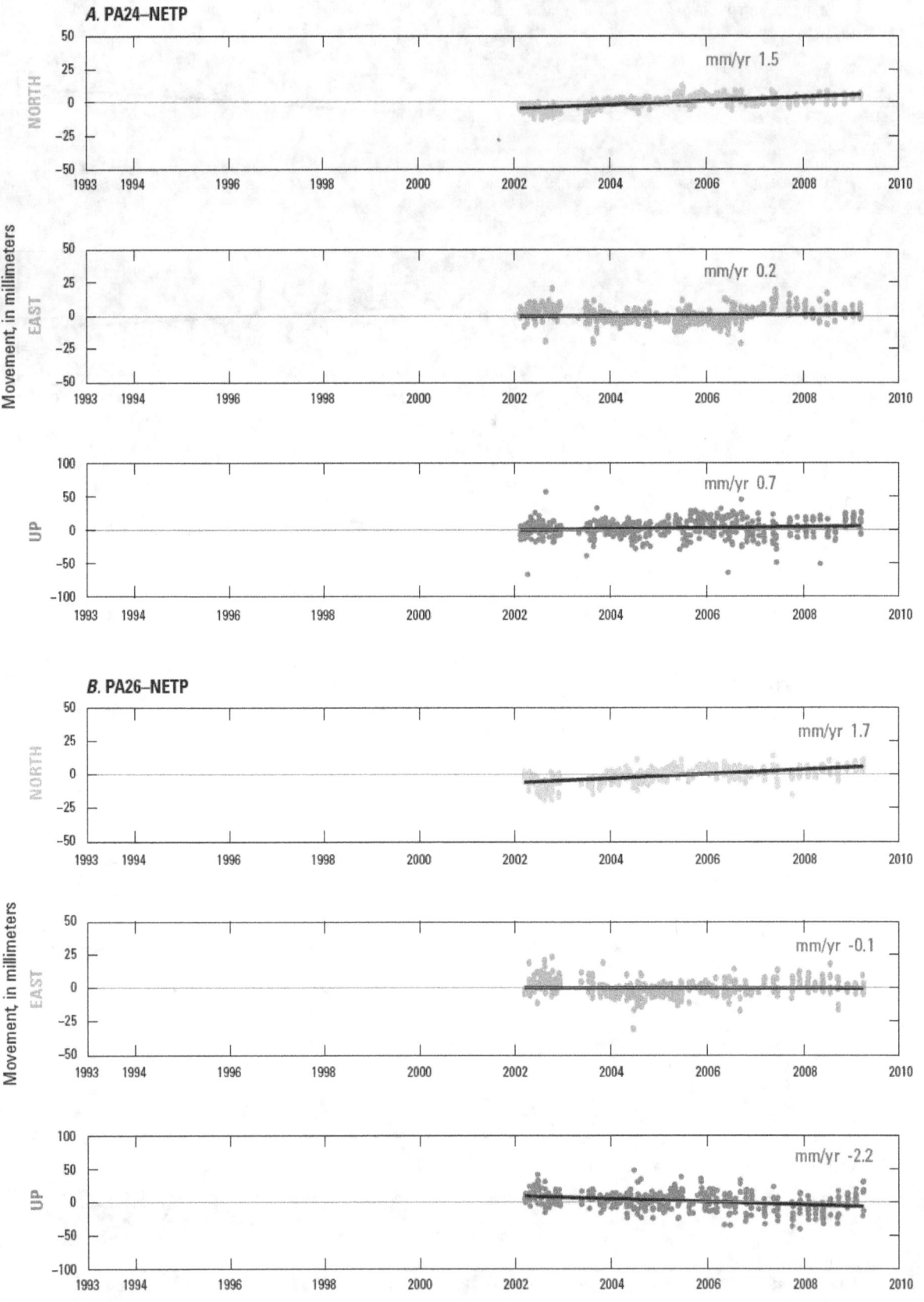

Figure 2.22.   Global Positioning System (GPS) time-series of relative movement with least-squares linear-regression lines depicting the directional movement in millimeters per year (mm/yr) between *A*, PAM 24 (PA24) and Northeast 2250 CORS ARP (NETP) Continuously Operating Reference Station (CORS) sites and *B*, PAM 26 (PA26) and Northeast 2250 CORS ARP (NETP) CORS sites.

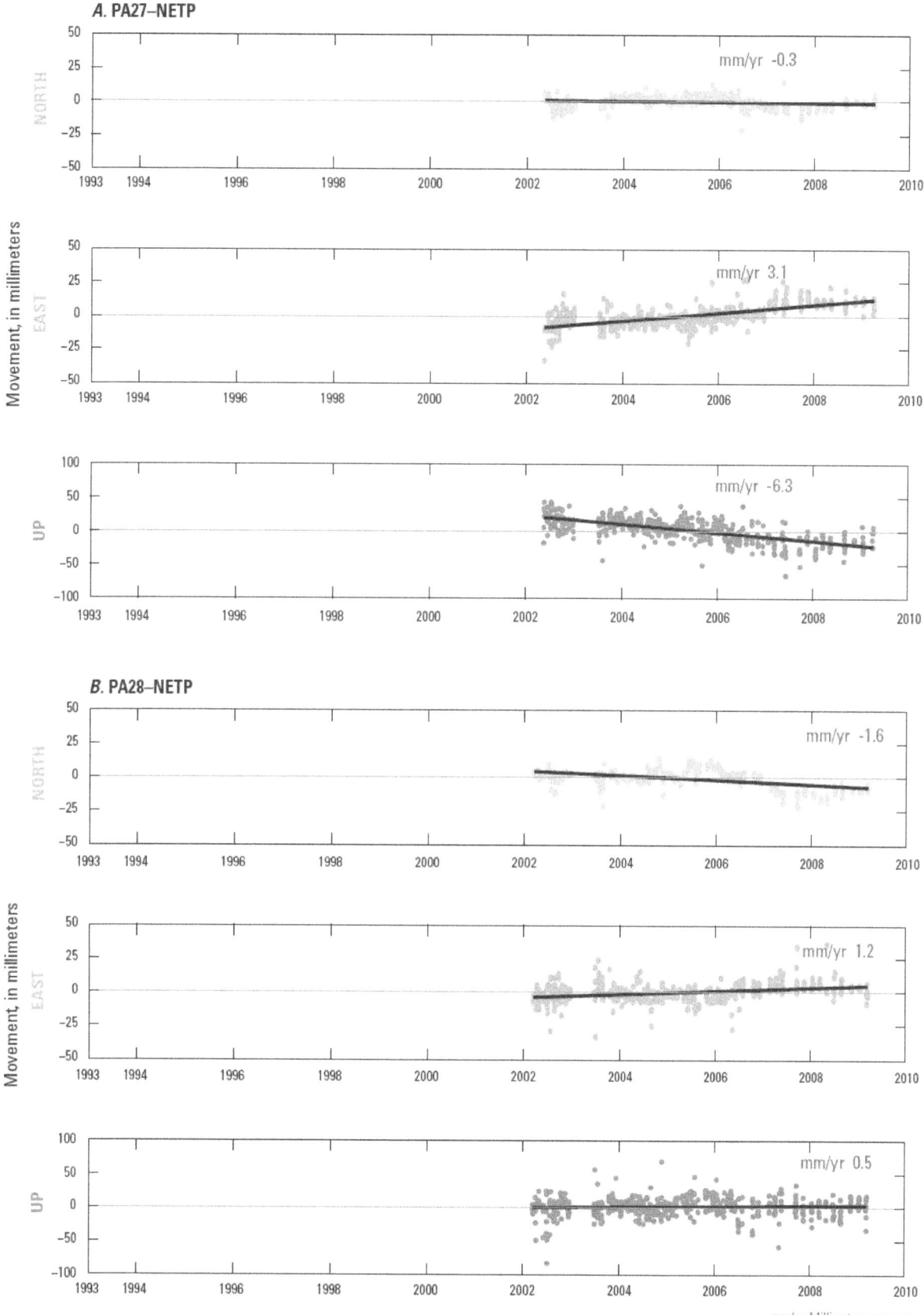

Figure 2.23.   Global Positioning System (GPS) time-series of relative movement with least-squares linear-regression lines depicting the directional movement in millimeters per year (mm/yr) between A, PAM 27 (PA27) and Northeast 2250 CORS ARP (NETP) Continuously Operating Reference Station (CORS) sites and B, PAM 28 (PA28) and Northeast 2250 CORS ARP (NETP) CORS sites.

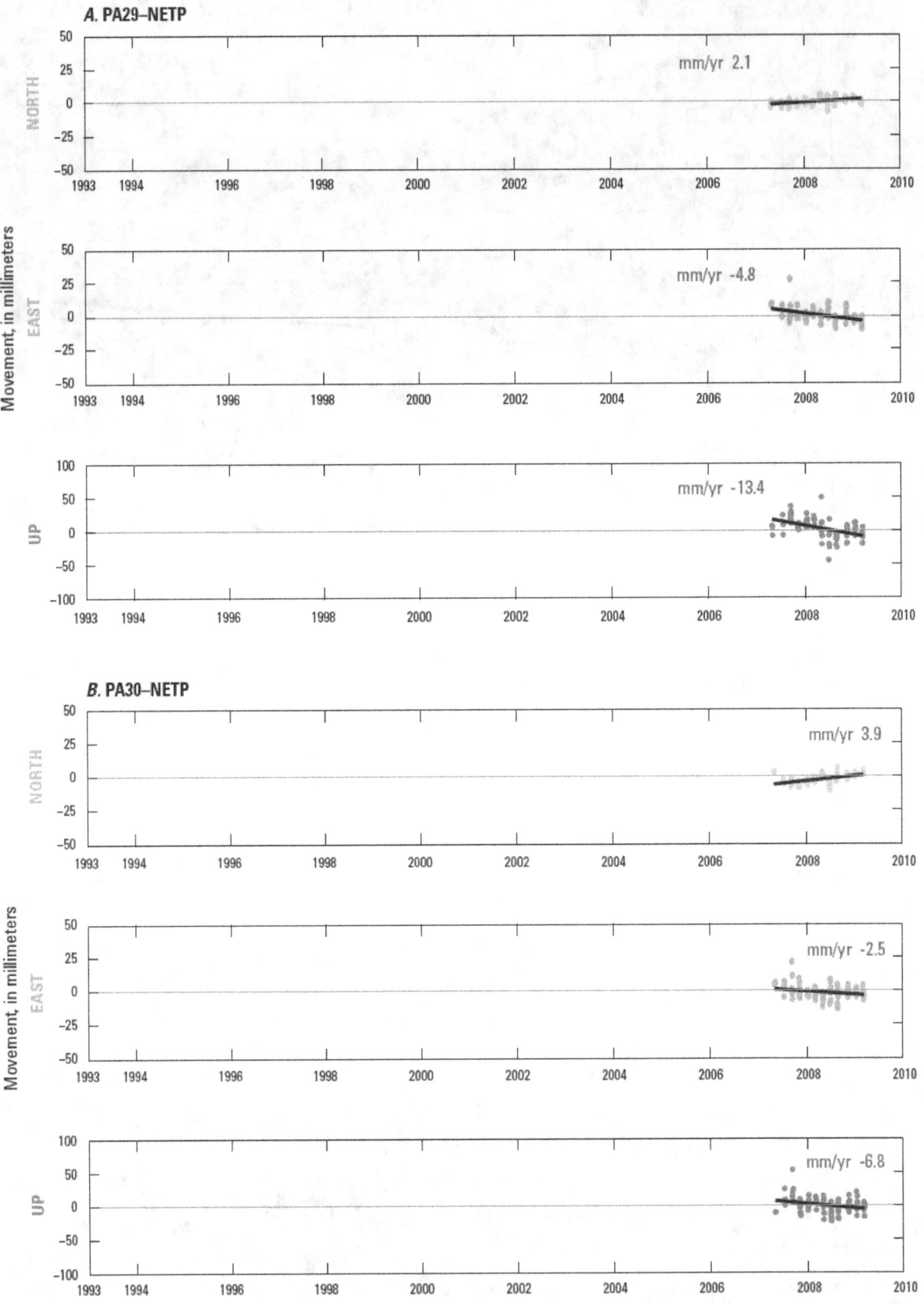

Figure 2.24.   Global Positioning System (GPS) time-series of relative movement with least-squares linear-regression lines depicting the directional movement in millimeters per year (mm/yr) between A, PAM 29 (PA29) and Northeast 2250 CORS ARP (NETP) Continuously Operating Reference Station (CORS) sites and B, PAM 30 (PA30) and Northeast 2250 CORS ARP (NETP) CORS sites.

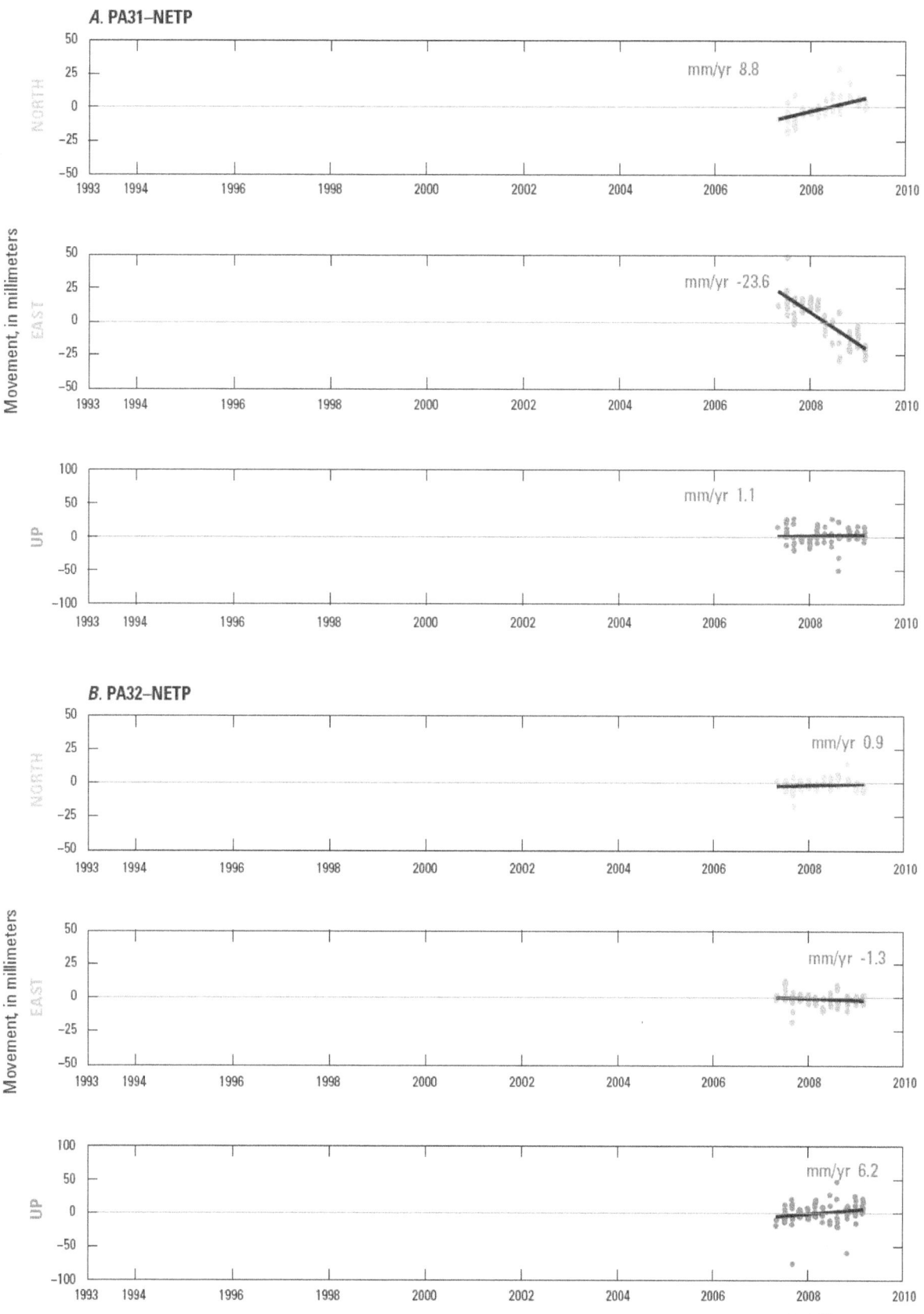

Figure 2.25. Global Positioning System (GPS) time-series of relative movement with least-squares linear-regression lines depicting the directional movement in millimeters per year (mm/yr) between *A*, PAM 31 (PA31) and Northeast 2250 CORS ARP (NETP) Continuously Operating Reference Station (CORS) sites and *B*, PAM 32 (PA32) and Northeast 2250 CORS ARP (NETP) CORS sites.

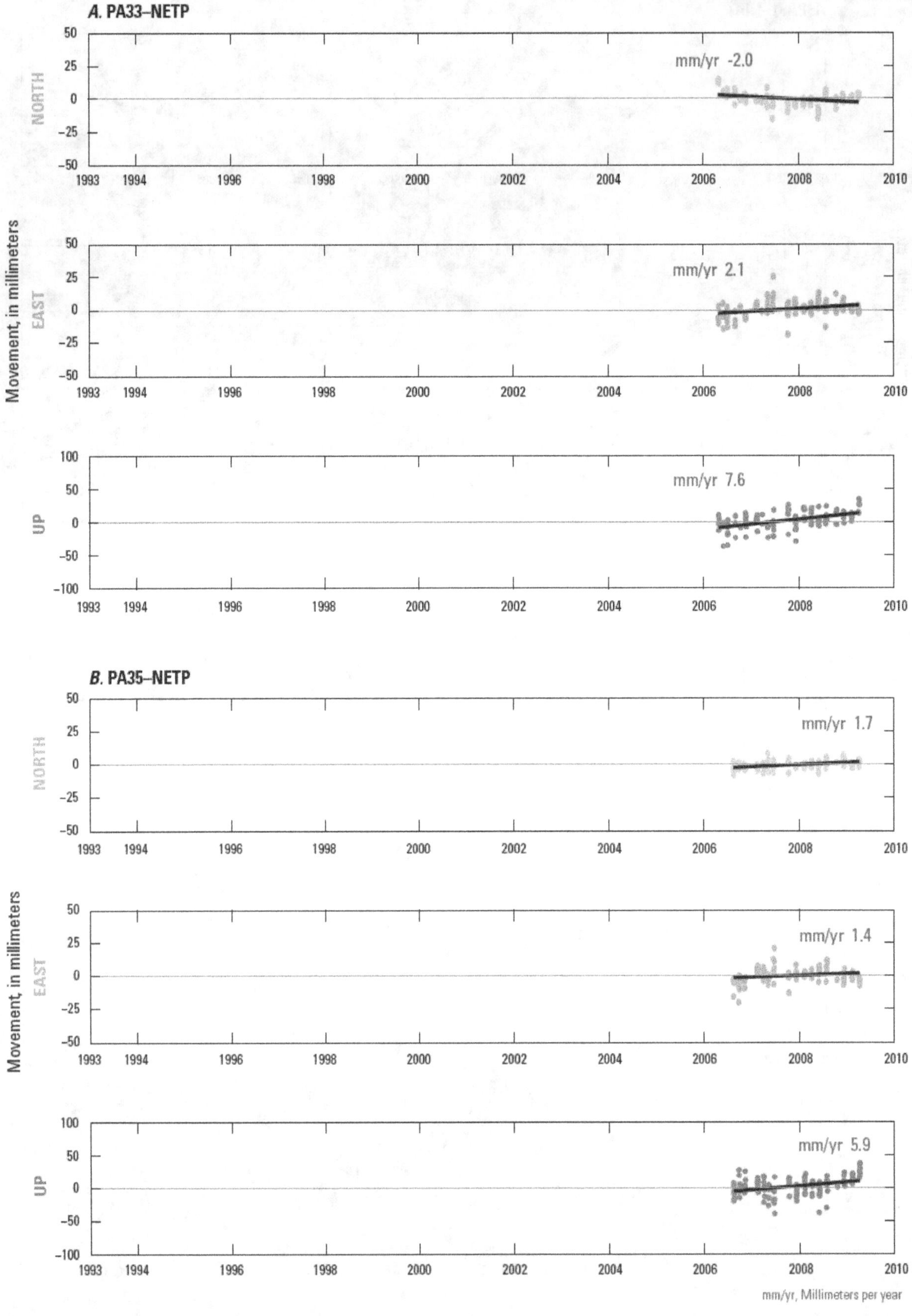

Figure 2.26.   Global Positioning System (GPS) time-series of relative movement with least-squares linear-regression lines depicting the directional movement in millimeters per year (mm/yr) between *A*, PAM 33 (PA33) and Northeast 2250 CORS ARP (NETP) Continuously Operating Reference Station (CORS) sites and *B*, PAM 35 (PA35) and Northeast 2250 CORS ARP (NETP) CORS sites.

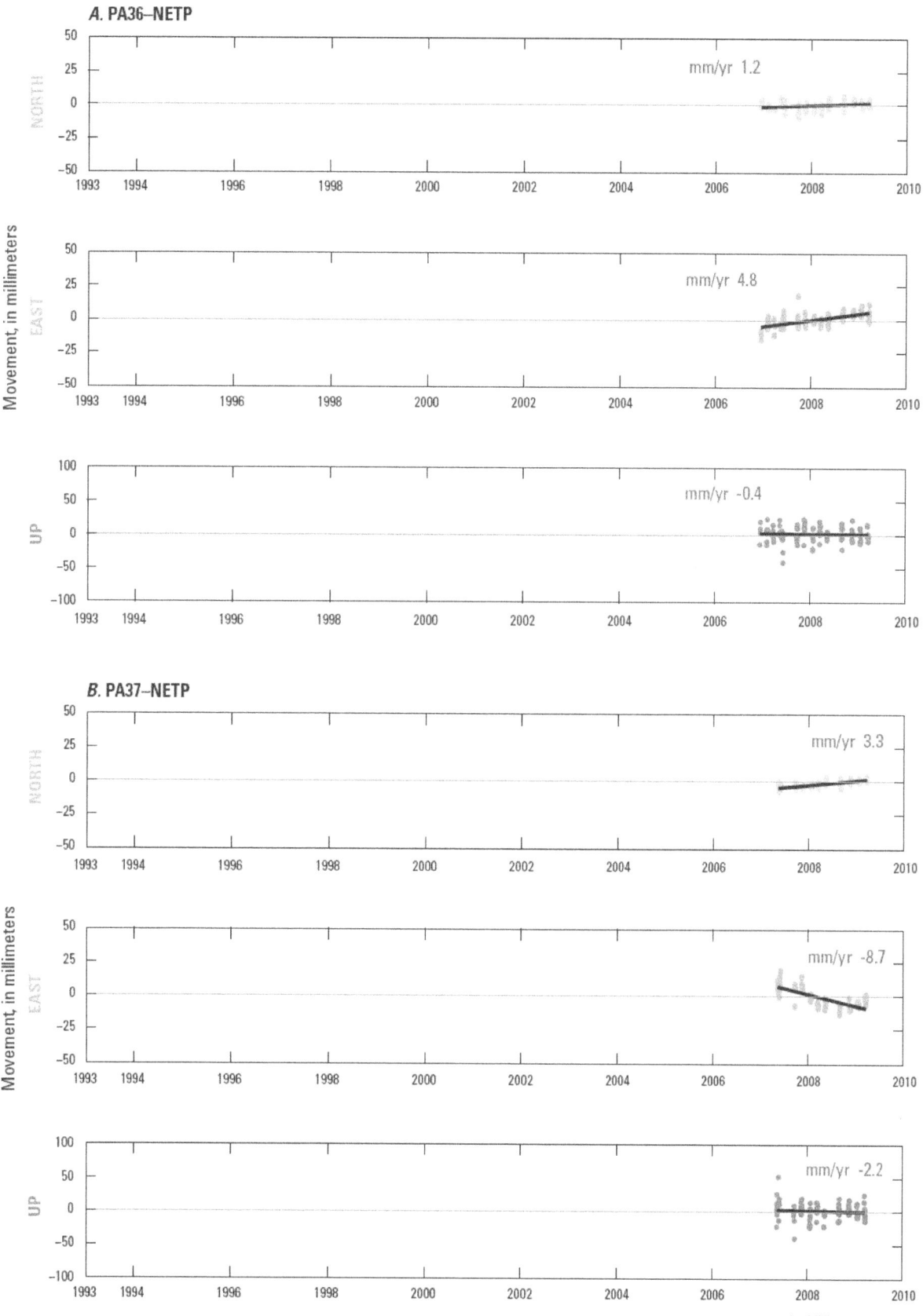

Figure 2.27.   Global Positioning System (GPS) time-series of relative movement with least-squares linear-regression lines depicting the directional movement in millimeters per year (mm/yr) between *A*, PAM 36 (PA36) and Northeast 2250 CORS ARP (NETP) Continuously Operating Reference Station (CORS) sites and *B*, PAM 37 (PA37) and Northeast 2250 CORS ARP (NETP) CORS sites.

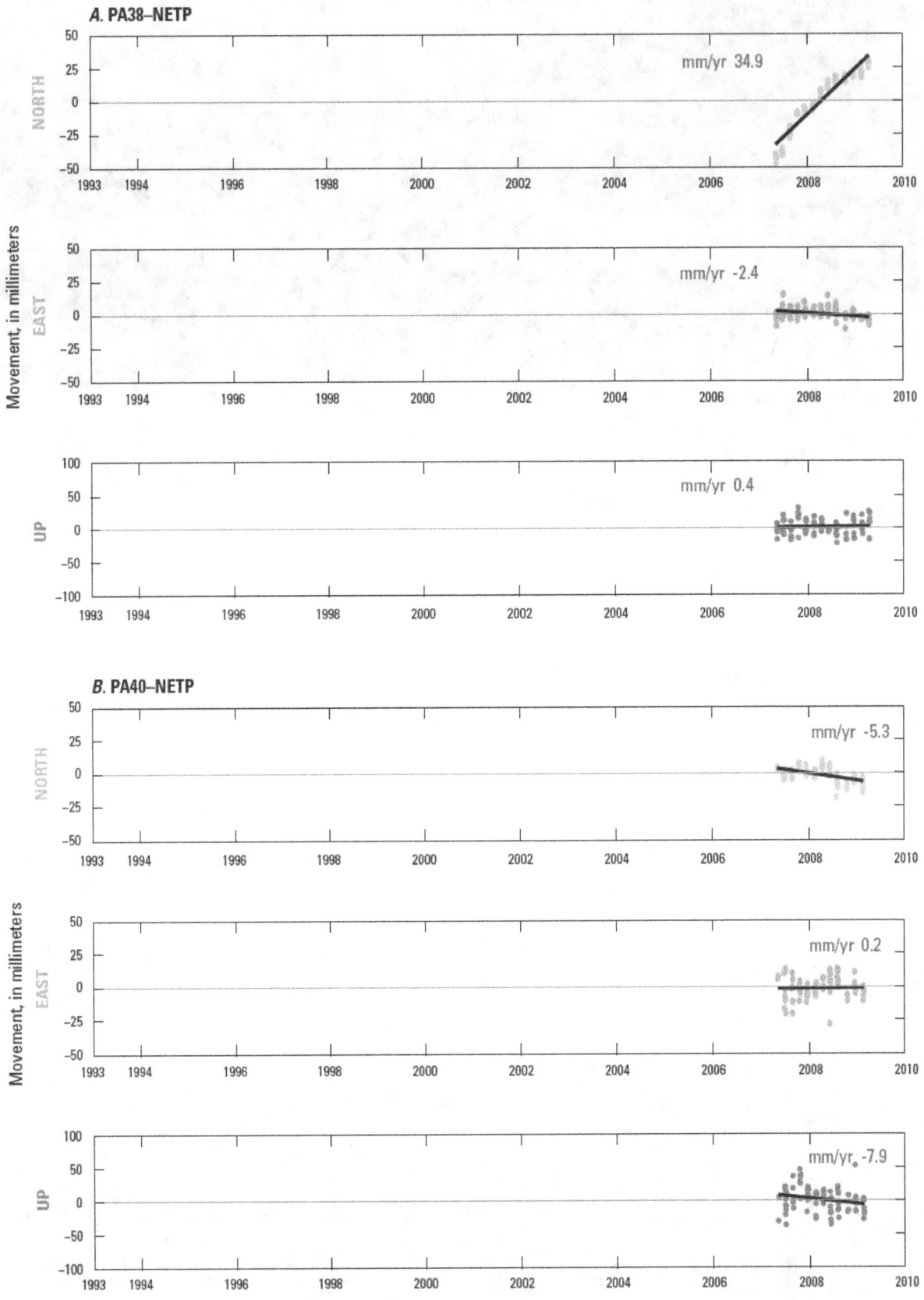

Figure 2.28. Global Positioning System (GPS) time-series of relative movement with least-squares linear-regression lines depicting the directional movement in millimeters per year (mm/yr) between A, PAM 38 (PA38) and Northeast 2250 CORS ARP (NETP) Continuously Operating Reference Station (CORS) sites and B, PAM 40 (PA40) and Northeast 2250 CORS ARP (NETP) CORS sites.

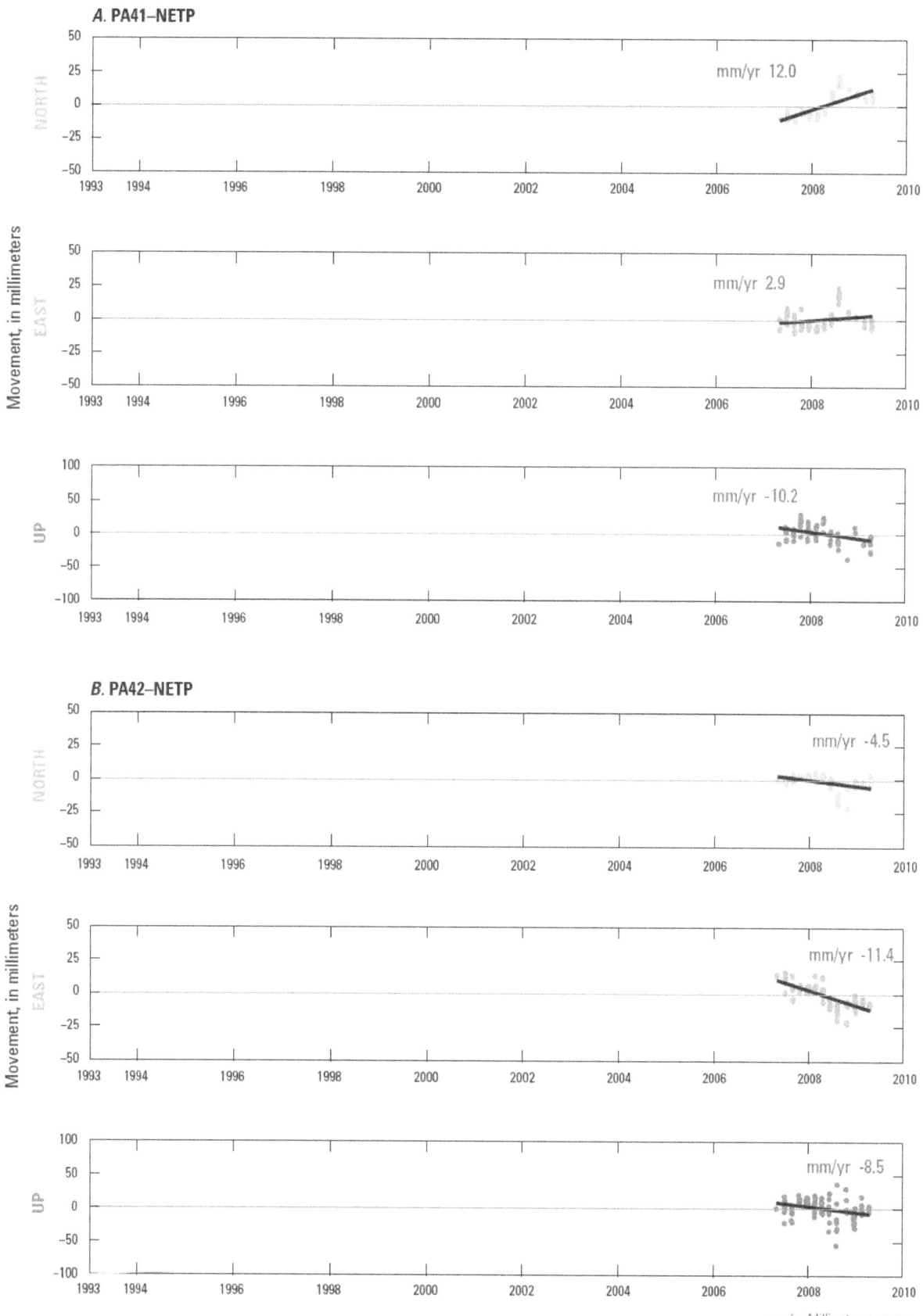

Figure 2.29.   Global Positioning System (GPS) time-series of relative movement with least-squares linear-regression lines depicting the directional movement in millimeters per year (mm/yr) between *A*, PAM 41 (PA41) and Northeast 2250 CORS ARP (NETP) Continuously Operating Reference Station (CORS) sites and *B*, PAM 42 (PA42) and Northeast 2250 CORS ARP (NETP) CORS sites.

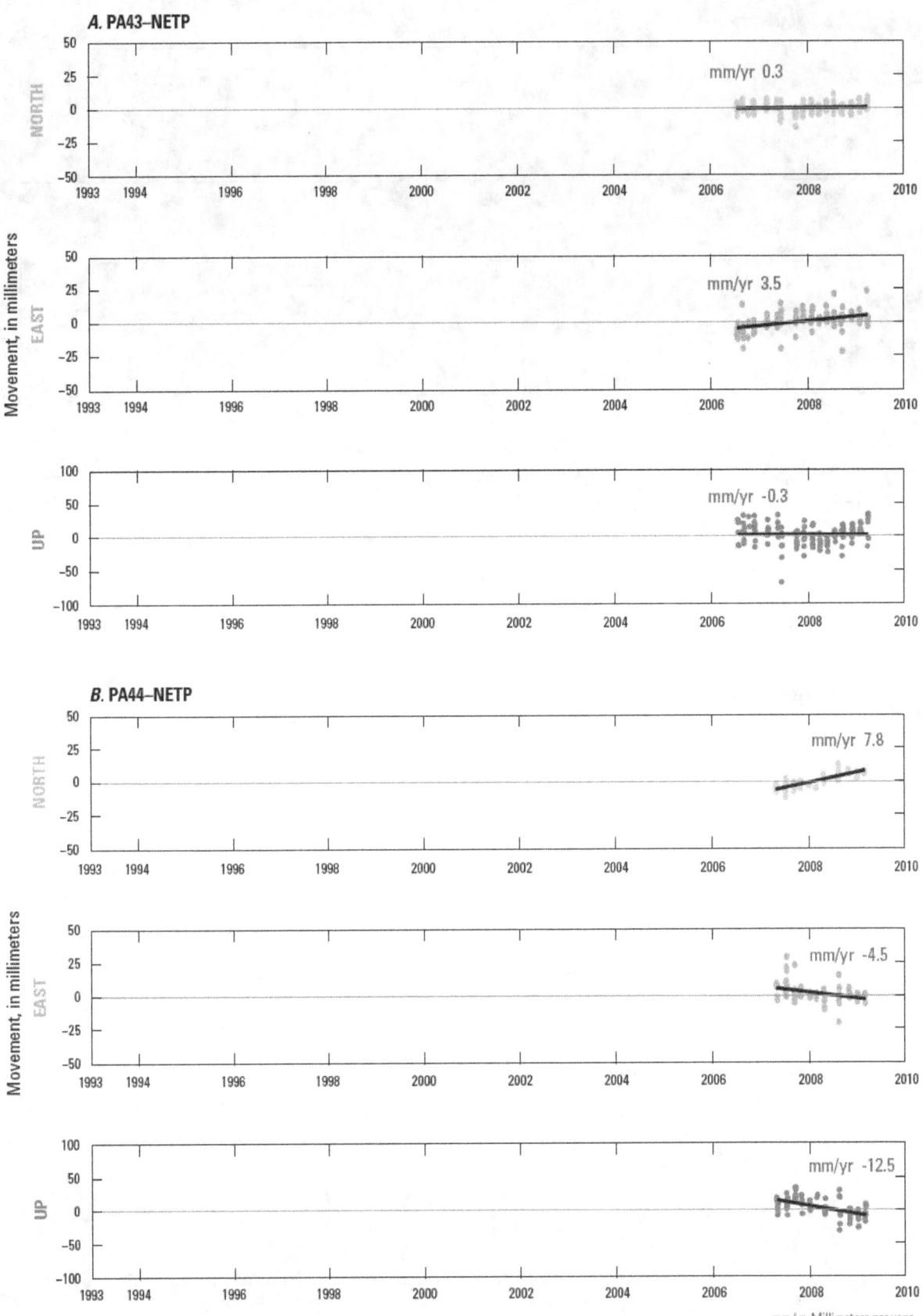

Figure 2.30.   Global Positioning System (GPS) time-series of relative movement with least-squares linear-regression lines depicting the directional movement in millimeters per year (mm/yr) between *A*, PAM 43 (PA43) and Northeast 2250 CORS ARP (NETP) Continuously Operating Reference Station (CORS) sites and *B*, PAM 44 (PA44) and Northeast 2250 CORS ARP (NETP) CORS sites.

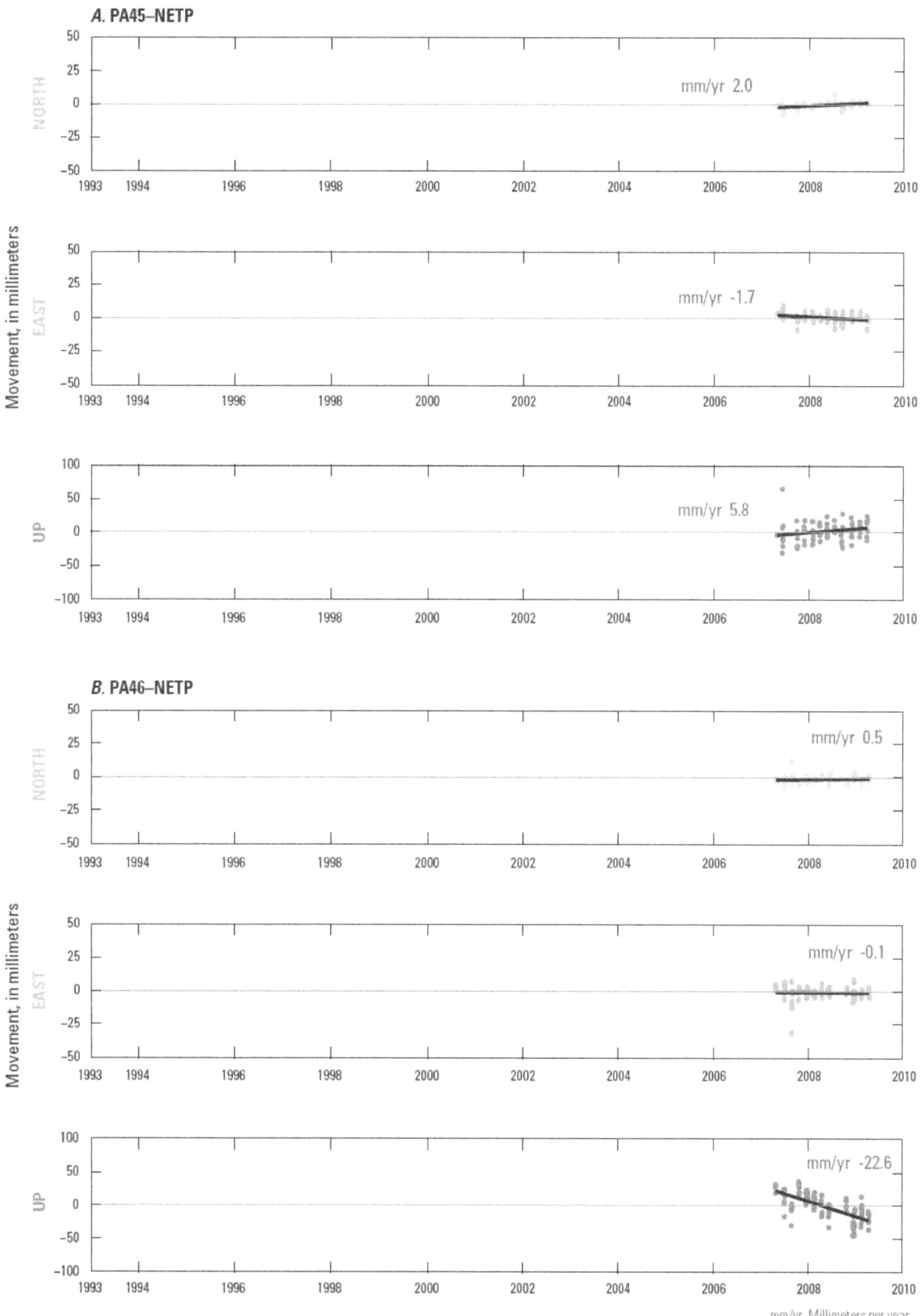

Figure 2.31.   Global Positioning System (GPS) time-series of relative movement with least-squares linear-regression lines depicting the directional movement in millimeters per year (mm/yr) between *A*, PAM 45 (PA45) and Northeast 2250 CORS ARP (NETP) Continuously Operating Reference Station (CORS) sites and *B*, PAM 46 (PA46) and Northeast 2250 CORS ARP (NETP) CORS sites.

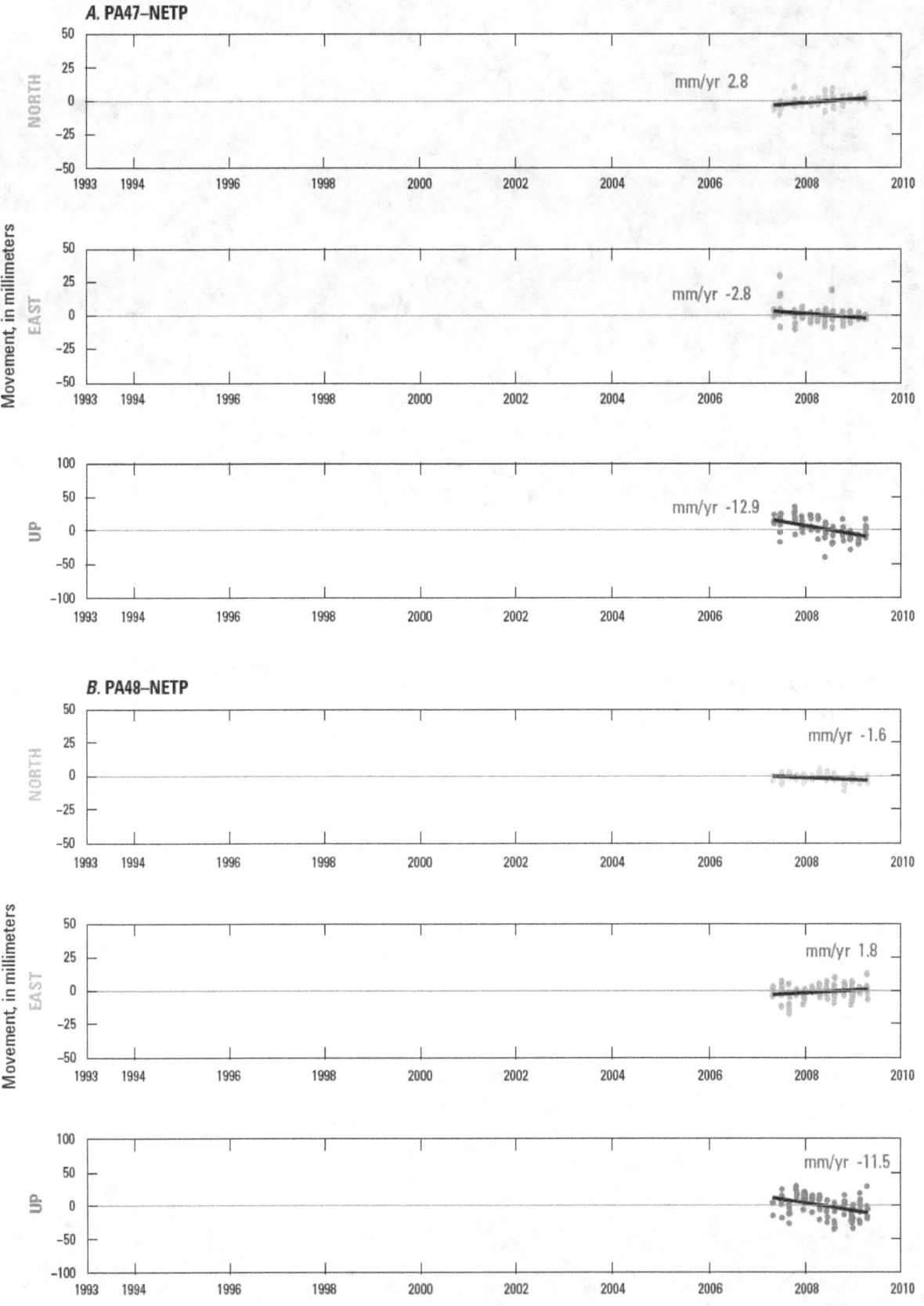

Figure 2.32. Global Positioning System (GPS) time-series of relative movement with least-squares linear-regression lines depicting the directional movement in millimeters per year (mm/yr) between *A*, PAM 47 (PA47) and Northeast 2250 CORS ARP (NETP) Continuously Operating Reference Station (CORS) sites and *B*, PAM 48 (PA48) and Northeast 2250 CORS ARP (NETP) CORS sites.

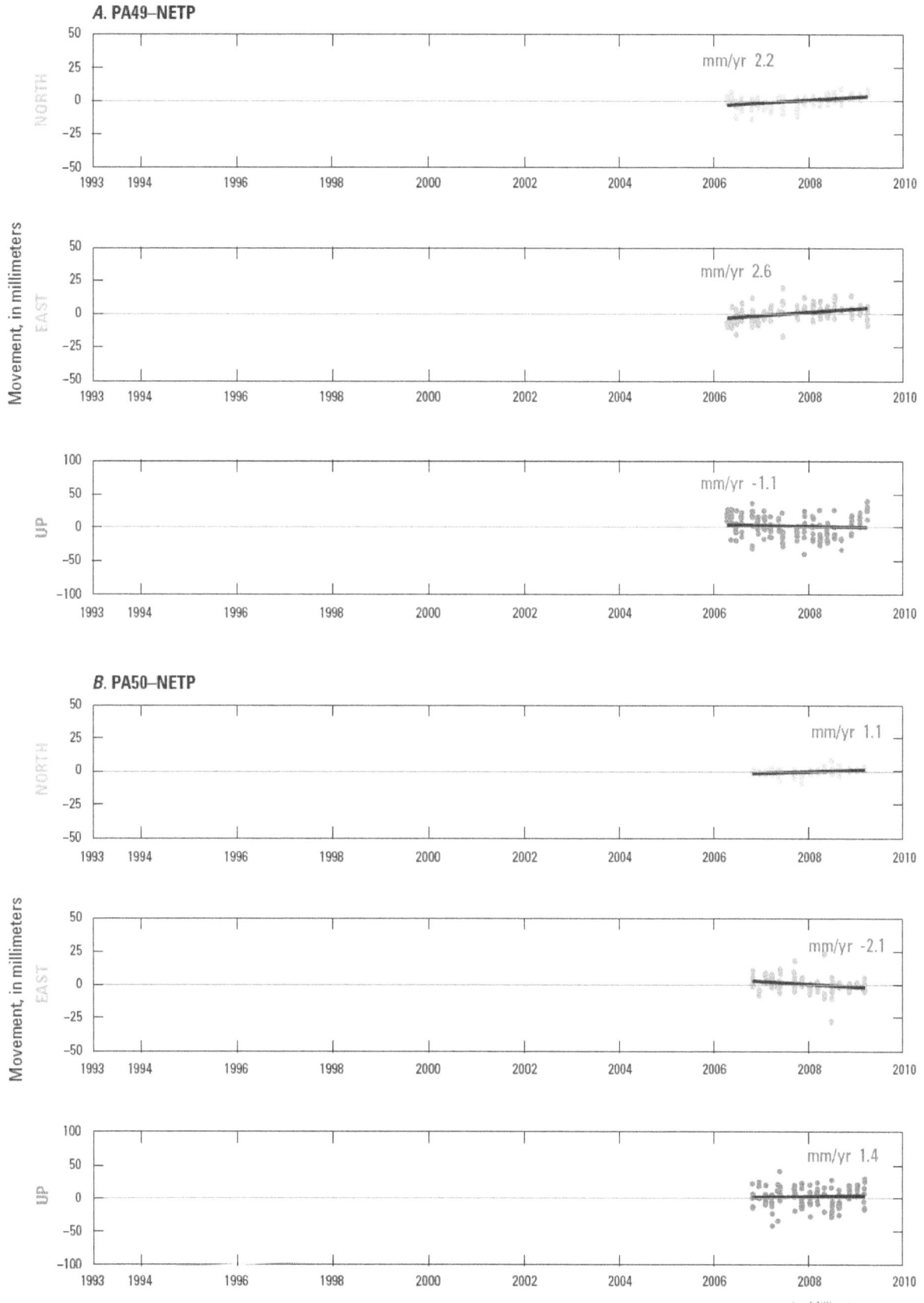

Figure 2.33. Global Positioning System (GPS) time-series of relative movement with least-squares linear-regression lines depicting the directional movement in millimeters per year (mm/yr) between *A*, PAM 49 (PA49) and Northeast 2250 CORS ARP (NETP) Continuously Operating Reference Station (CORS) sites and *B*, PAM 50 (PA40) and Northeast 2250 CORS ARP (NETP) CORS sites.

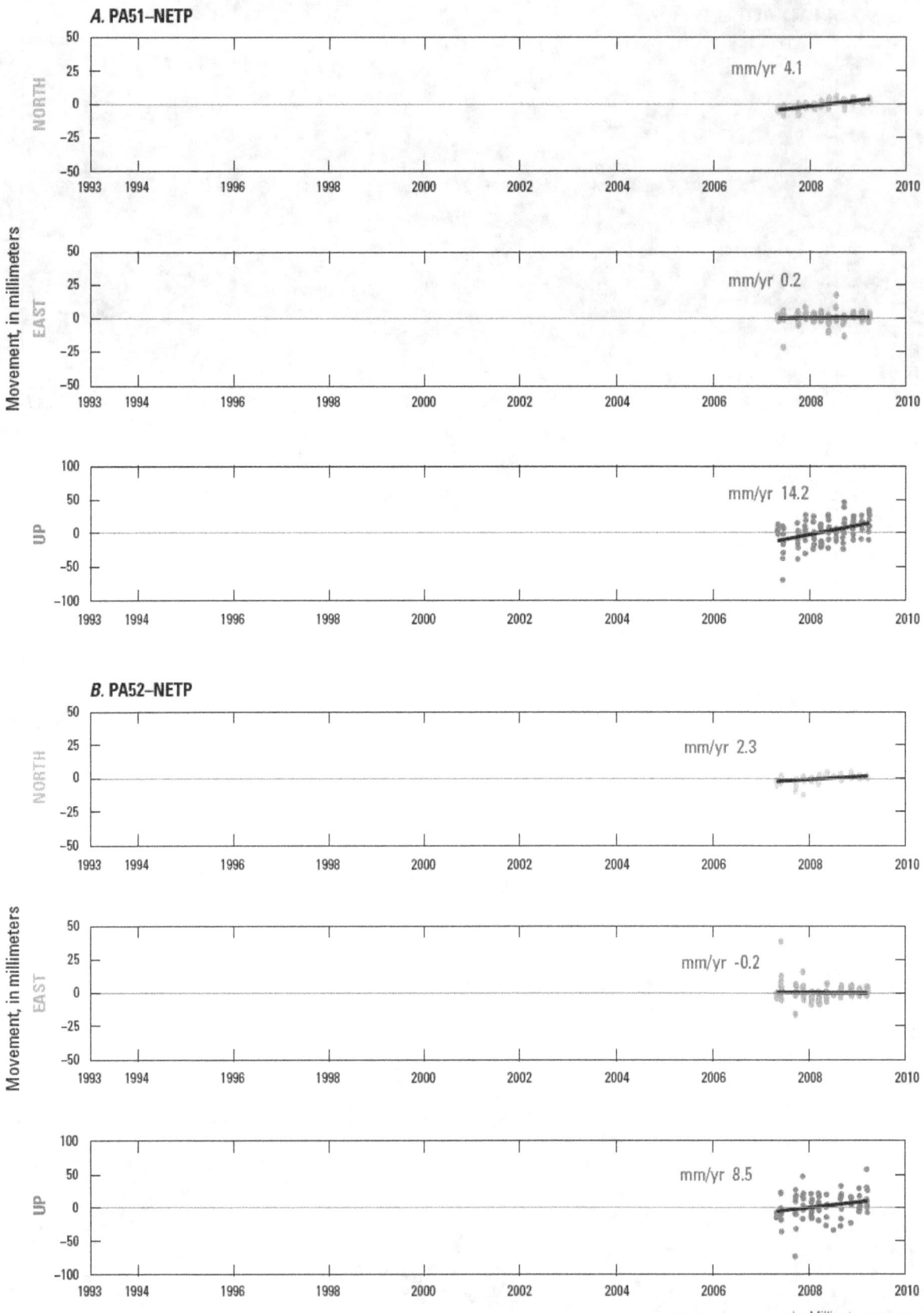

Figure 2.34.   Global Positioning System (GPS) time-series of relative movement with least-squares linear-regression lines depicting the directional movement in millimeters per year (mm/yr) between *A*, PAM 51 (PA51) and Northeast 2250 CORS ARP (NETP) Continuously Operating Reference Station (CORS) sites and *B*, PAM 52 (PA52) and Northeast 2250 CORS ARP (NETP) CORS sites.

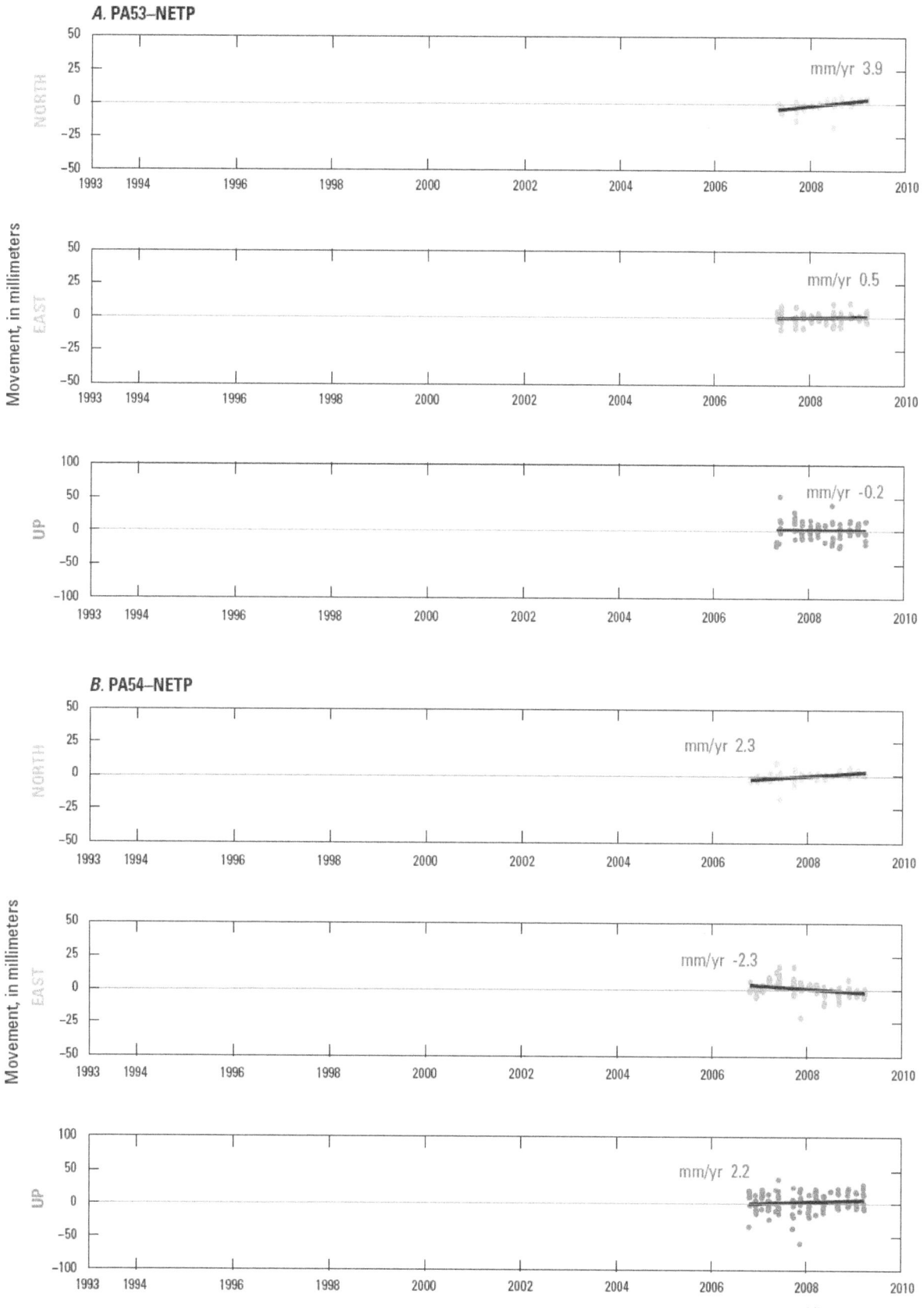

Figure 2.35. Global Positioning System (GPS) time-series of relative movement with least-squares linear-regression lines depicting the directional movement in millimeters per year (mm/yr) between A, PAM 53 (PA53) and Northeast 2250 CORS ARP (NETP) Continuously Operating Reference Station (CORS) sites and B, PAM 54 (PA54) and Northeast 2250 CORS ARP (NETP) CORS sites.

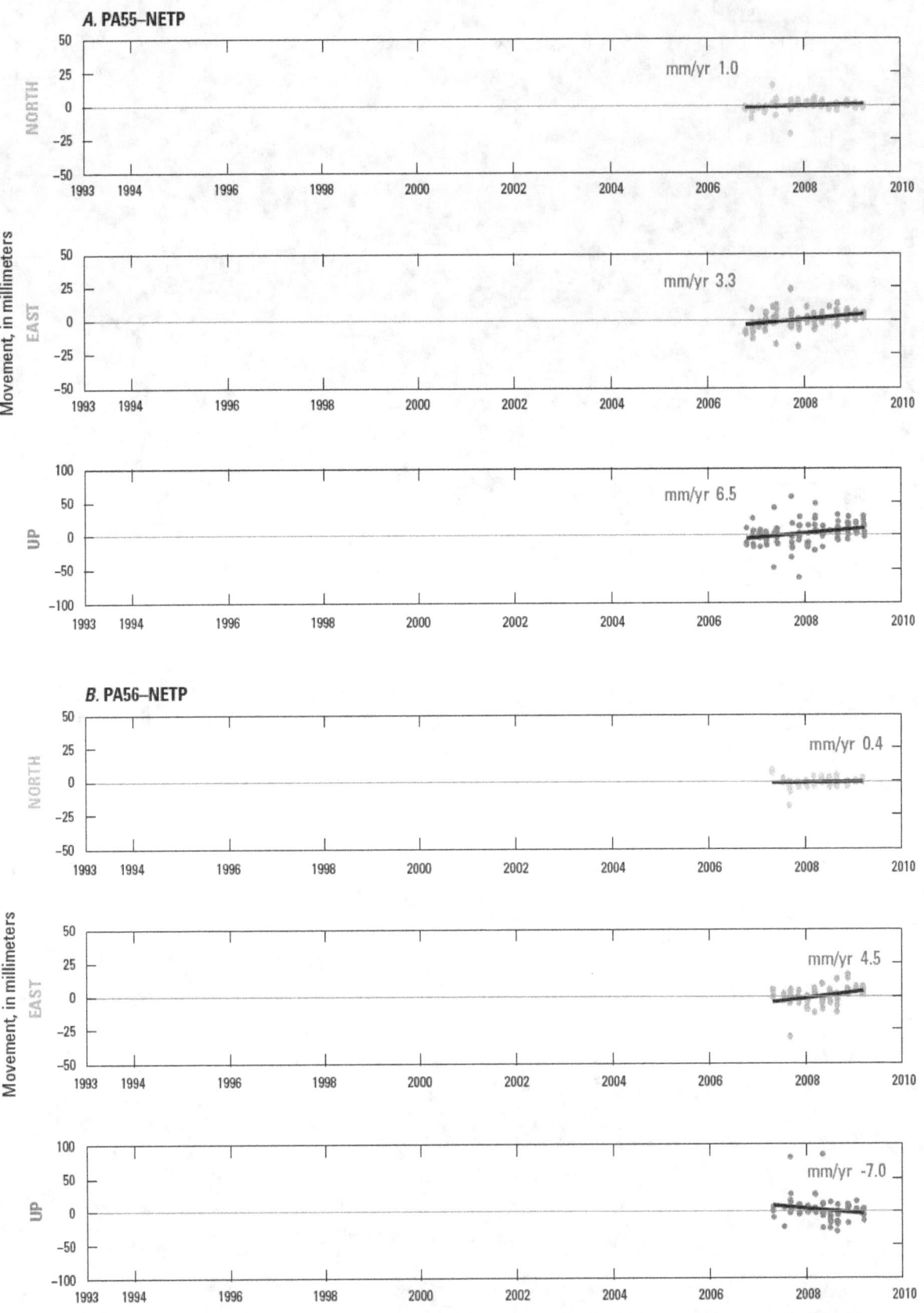

mm/yr, Millimeters per year

Figure 2.36.   Global Positioning System (GPS) time-series of relative movement with least-squares linear-regression lines depicting the directional movement in millimeters per year (mm/yr) between *A*, PAM 55 (PA55) and Northeast 2250 CORS ARP (NETP) Continuously Operating Reference Station (CORS) sites and *B*, PAM 56 (PA56) and Northeast 2250 CORS ARP (NETP) CORS sites.

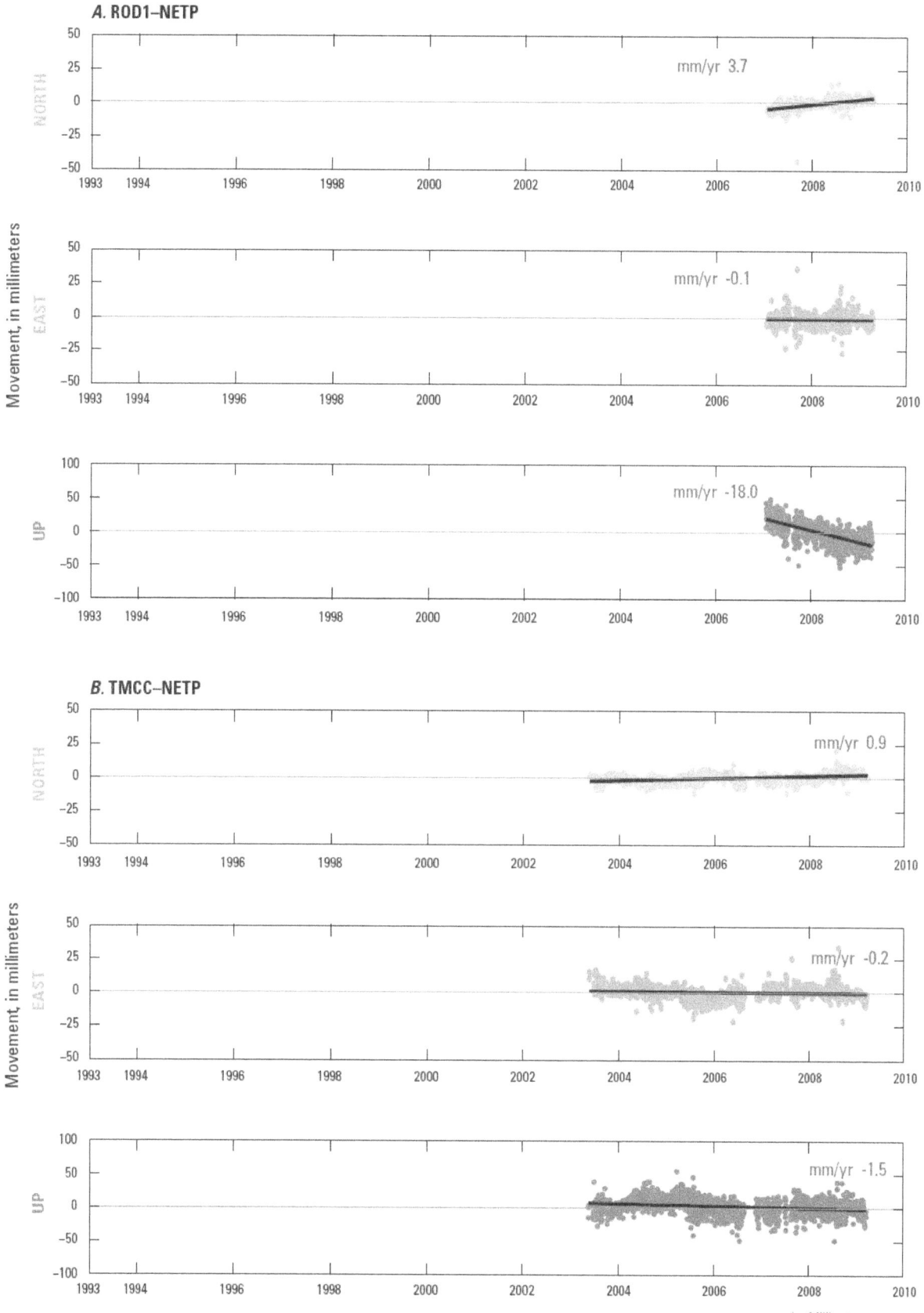

Figure 2.37. Global Positioning System (GPS) time-series of relative movement with least-squares linear-regression lines depicting the directional movement in millimeters per year (mm/yr) between A, Spring (ROD1) and Northeast 2250 CORS ARP (NETP) Continuously Operating Reference Station (CORS) sites and B, TMCC and Northeast 2250 CORS ARP (NETP) CORS sites.

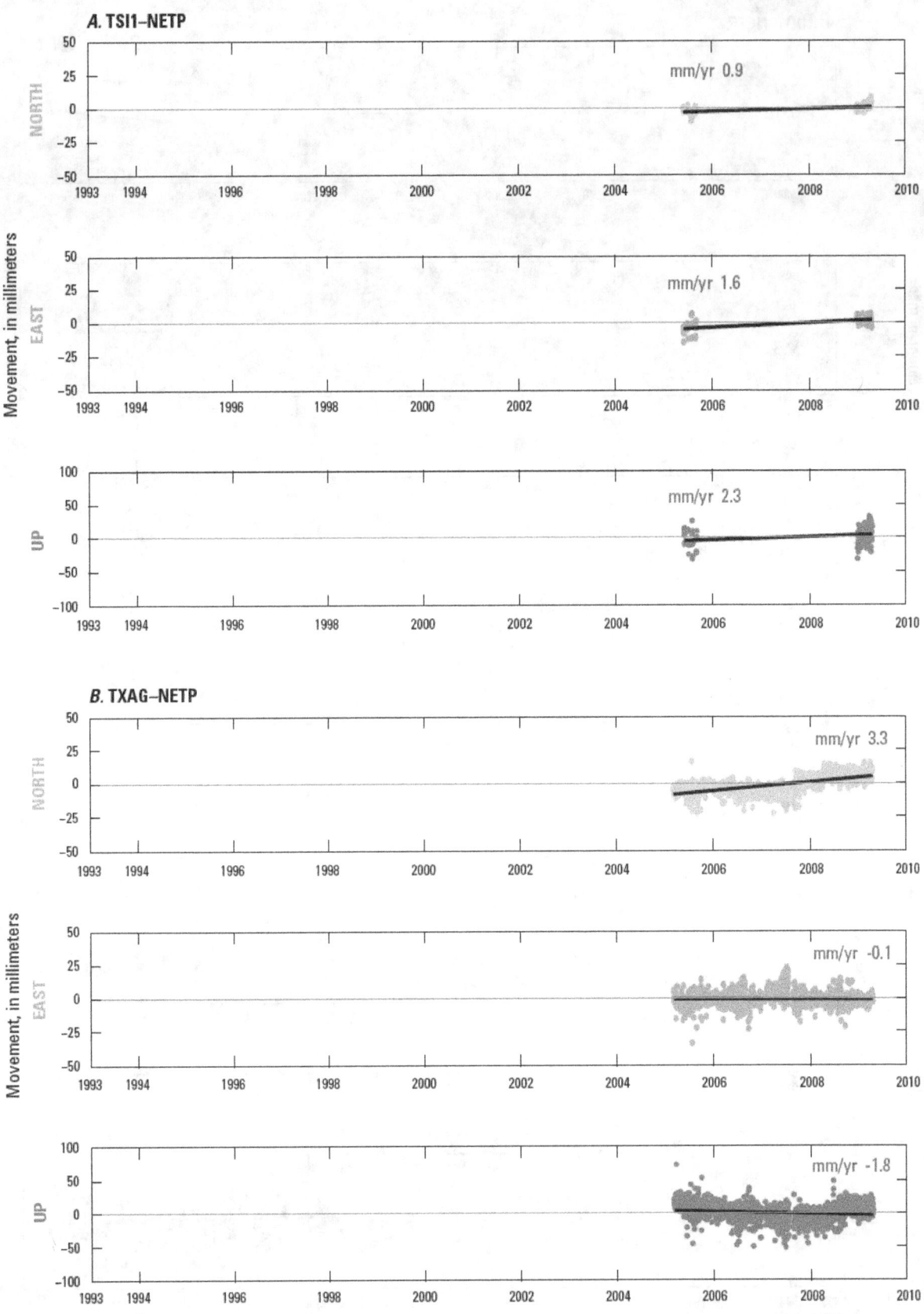

Figure 2.38.   Global Positioning System (GPS) time-series of relative movement with least-squares linear-regression lines depicting the directional movement in millimeters per year (mm/yr) between *A*, Deerpark COOP (TSI1) and Northeast 2250 CORS ARP (NETP) Continuously Operating Reference Station (CORS) sites and *B*, Angleton (TXAG) and Northeast 2250 CORS ARP (NETP) CORS sites.

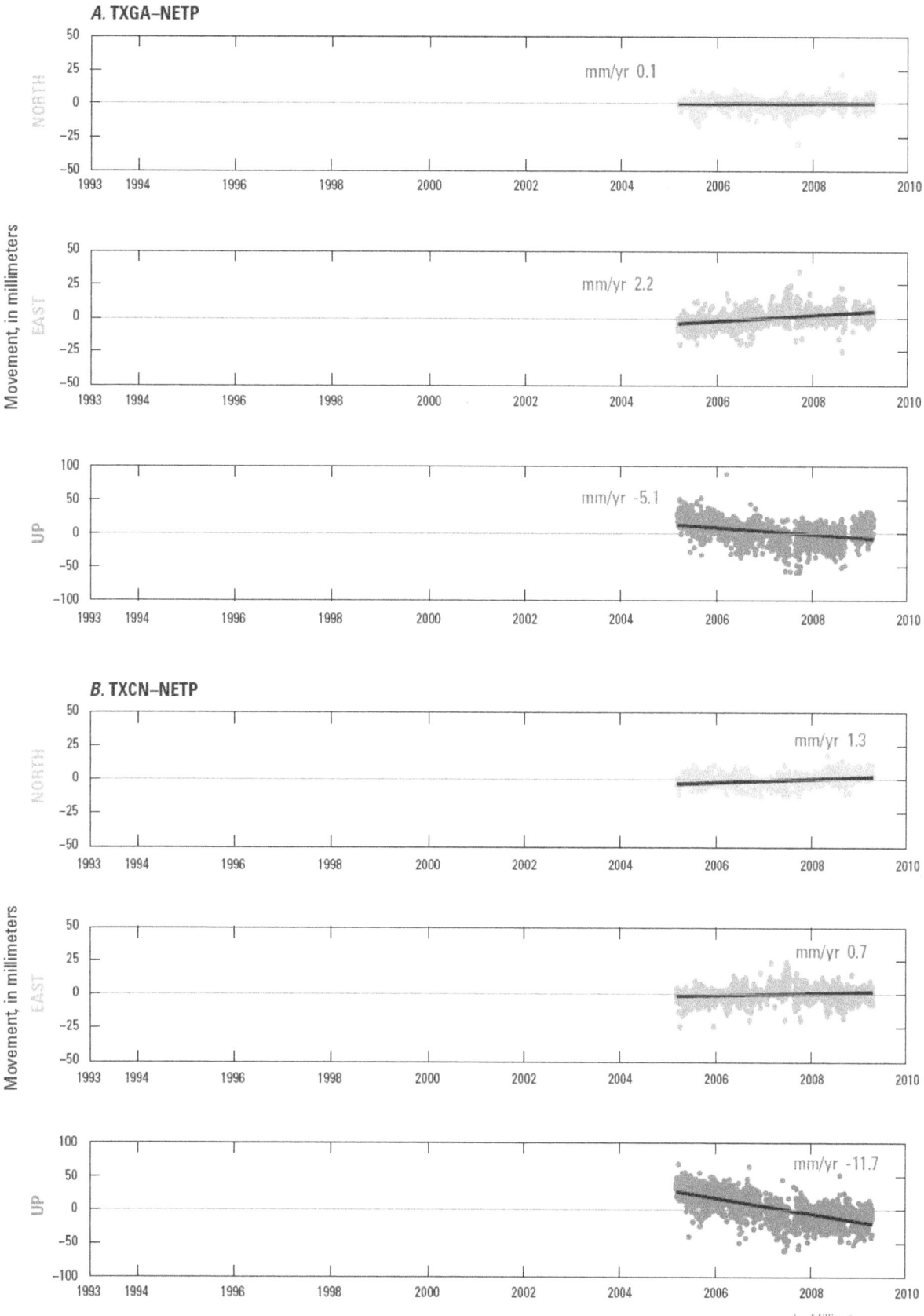

Figure 2.39. Global Positioning System (GPS) time-series of relative movement with least-squares linear-regression lines depicting the directional movement in millimeters per year (mm/yr) between A, Galveston (TXGA) and Northeast 2250 CORS ARP (NETP) Continuously Operating Reference Station (CORS) sites and B, Conroe (TXCN) and Northeast 2250 CORS ARP (NETP) CORS sites.

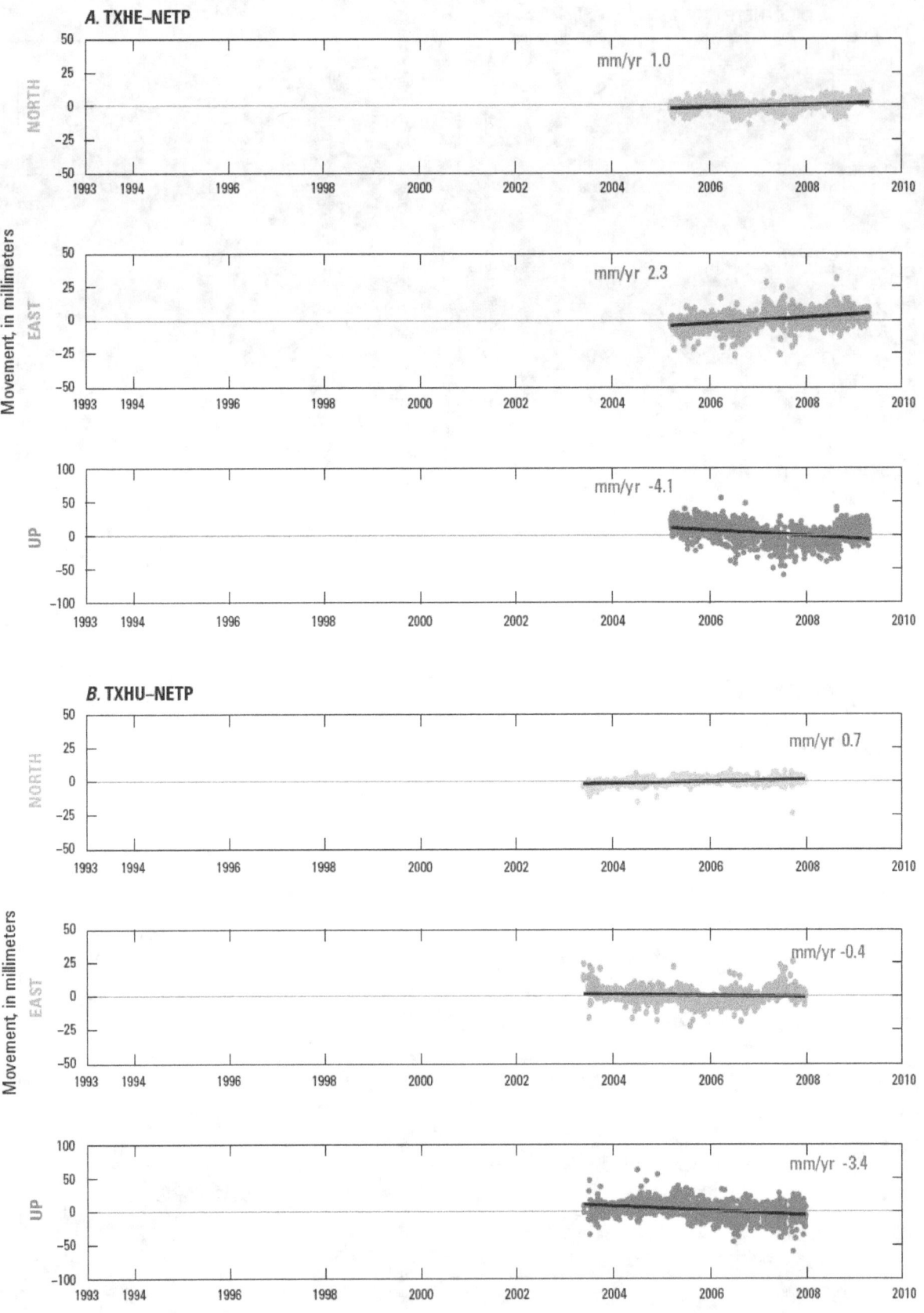

Figure 2.40.   Global Positioning System (GPS) time-series of relative movement with least-squares linear-regression lines depicting the directional movement in millimeters per year (mm/yr) between *A*, Hempstead (TXHE) and Northeast 2250 CORS ARP (NETP) Continuously Operating Reference Station (CORS) sites and *B*, Houston RRP2 (TXHU) and Northeast 2250 CORS ARP (NETP) CORS sites.

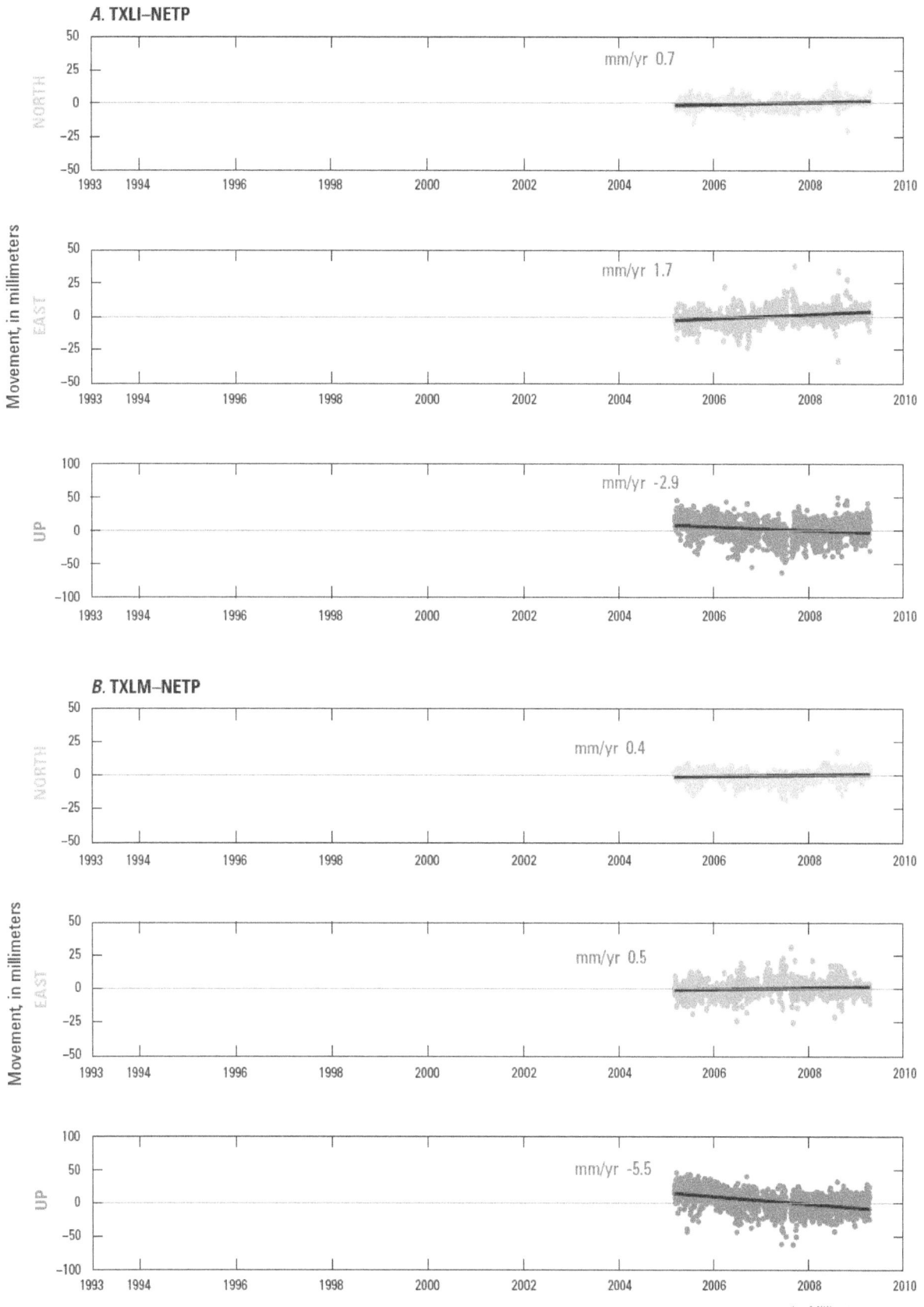

Figure 2.41.   Global Positioning System (GPS) time-series of relative movement with least-squares linear-regression lines depicting the directional movement in millimeters per year (mm/yr) between *A*, Liberty (TXLI) and Northeast 2250 CORS ARP (NETP) Continuously Operating Reference Station (CORS) sites and *B*, LaMarque (TXLM) and Northeast 2250 CORS ARP (NETP) CORS sites.

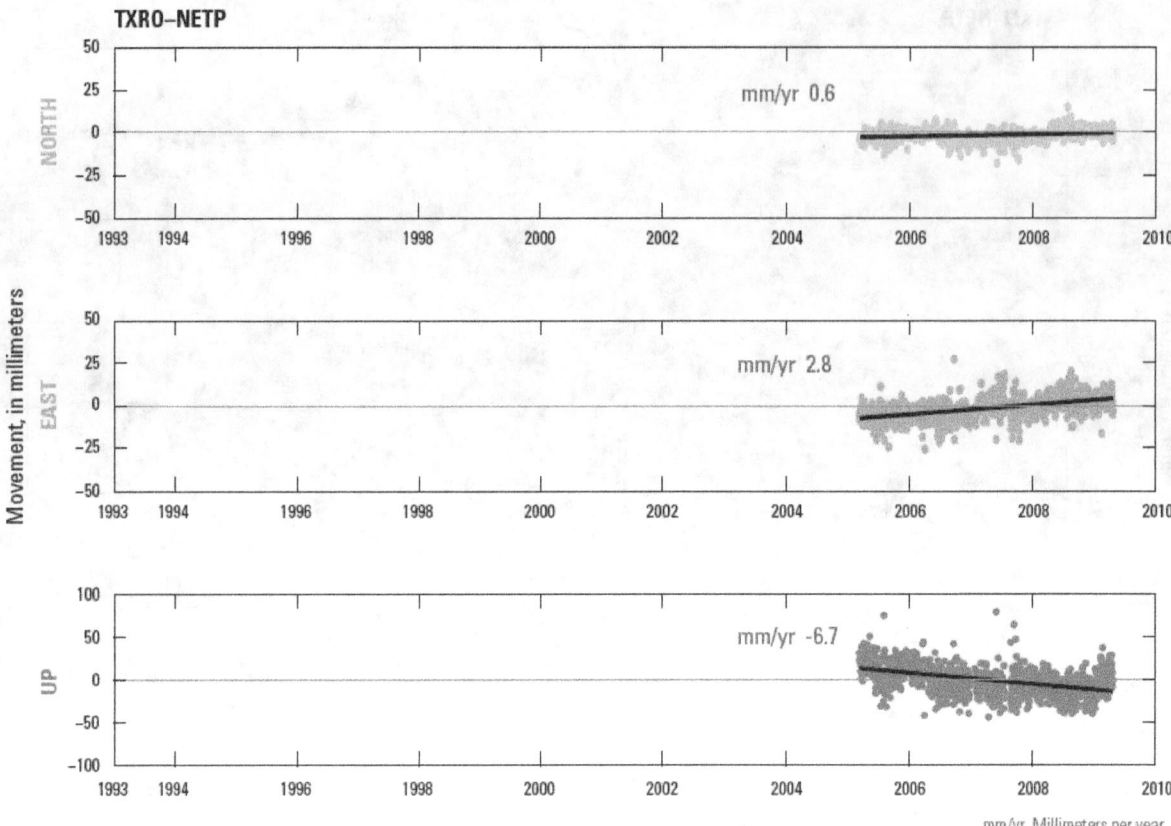

Figure 2.42.   Global Positioning System (GPS) time-series of relative movement with least-squares linear-regression lines depicting the directional movement in millimeters per year (mm/yr) between Rosenberg (TXRO) and Northeast 2250 CORS ARP (NETP) Continuously Operating Reference Station (CORS) sites.

Publishing support provided by
Lafayette Publishing Service Center

Information regarding water resources in Texas is available at
http://tx.usgs.gov/

www.ingramcontent.com/pod-product-compliance
Lightning Source LLC
Chambersburg PA
CBHW081547170526
45166CB00009B/2612